# Intelligent Video Surveillance - New Perspectives

*Edited by Pier Luigi Mazzeo*

Published in London, United Kingdom

Intelligent Video Surveillance - New Perspectives
http://dx.doi.org/10.5772/intechopen.100777
Edited by Pier Luigi Mazzeo

Contributors
İbrahim Delibaşoğlu, Ashwin Yadav, Kamal Jain, Akshay Pandey, Joydeep Majumdar, Rohit Sahay, Yu Chen, Deeraj Nagothu, Nihal Poredi, Emily M. Hand, Sara R. Davis, Muhammad H. Hamza El-Saba, Jian Liang, Liang An, Majid Mirbod, Kaoru Sumi, Nitchan Jianwattanapaisarn, Akira Utsumi, Pier Luigi Mazzeo



First published in London, United Kingdom, 2023 by IntechOpen
IntechOpen is the global imprint of INTECHOPEN LIMITED, registered in England and Wales, registration number: 11086078, 5 Princes Gate Court, London, SW7 2QJ, United Kingdom

British Library Cataloguing-in-Publication Data
A catalogue record for this book is available from the British Library

Additional hard and PDF copies can be obtained from orders@intechopen.com

Intelligent Video Surveillance - New Perspectives
Edited by Pier Luigi Mazzeo
p. cm.
Print ISBN 978-1-80356-341-1
Online ISBN 978-1-80356-342-8
eBook (PDF) ISBN 978-1-80356-343-5

# Meet the editor

Pier Luigi Mazzeo received his MSc in computer science from the University of Salento in 2001. He joined the Italian National Research Council (CNR) as a researcher in 2002, and since then he has been working on several artificial intelligence and computer vision research topics. The projects in which he is currently involved include algorithms for video object tracking, face detection and recognition, facial expression recognition, deep neural networks (DNN) and machine learning. He has authored and co-authored a total of 100 publications, including more than 20 papers published in international journals and book chapters. He has also co-authored five national and international patents. He acts as a reviewer for several international journals and for some book publishers. Since 2004, Pier Luigi has been regularly invited to take part in the scientific committees of national and international conferences.

# Contents

# Preface

In recent years, scientific and technological advances in areas such as computer vision, artificial intelligence and connectivity between systems and processes have created the conditions for the development and spread of intelligent video surveillance (IVS), a new field of research that has given rise to a considerable number of new studies exploring its uses and potential. With more and more cameras installed and the increasing complexity of the required algorithms, in-house self-contained video surveillance systems have become a necessity for most institutions and companies. New achievements in the field of IVS help not only storage space in the cloud (necessary for handling large amounts of video data), but also infrastructures and computational power to be distributed.

This book introduces the main features of IVS, and provides a case study where research-lab computer vision algorithms are integrated into an IVS solution. The lessons learned and some future directions for this topic are also discussed.

The book has two sections: the first explores different applications of intelligent video surveillance, and the second investigates new methodologies for introducing intelligence in video surveillance solutions.

The Editor warmly thanks all the chapter authors and hopes that the publication of this interesting book contributes to designing a future where technologies enable people to live comfortably, healthily, and mainly in peace.

**Pier Luigi Mazzeo**
National Research Council of Italy (CNR),
Institute of Applied Sciences and Intelligent Systems (ISASI)
Lecce, Italy

Section 1

# Context-Driven Applications

**Chapter 1**

# Introductory Chapter: Intelligent Video Surveillance – What's Next?

*Pier Luigi Mazzeo*

## 1. Introduction

Most of the biggest urban areas are speedily translating into smart cities with structures able to handle huge quantity of data shaped by the Internet of Things apparatus for intelligent analysis. Decreasing human contribution and improving people's life quality are two main objectives that smart cities proposed to reach. A concrete research field and its applications that satisfy both of these principles is smart video surveillance [1].

Intelligent video surveillance systems aim to analyze the observed scenario using machine learning, computer vision, and data analytics in order to minimize or completely eradicate human contribution.

The request for such intelligent security systems qualified for recognizing both natural emergencies, such as fire, floods, earthquakes, and human-made emergencies, such as violence, traffic accidents, and weapon threats, is growing solidly [2].

Intelligent video surveillance systems are usually adopted in different contexts, spreading from public areas and infrastructures to commercial buildings. They are often used for a double scope: i) real-time monitoring of physical estates and areas and ii) for reviewing collected video information to estimate security indicators and plan safety measures, consequently.

In the last decades, intelligent video surveillance systems are deeply employed in the public and security sectors, but now a significant interest in these topics has quickly been raised by other stakeholders. This interest has been caused by the constant increase in crime rates and security national and international threats, which are conducting incredible growth in the market of video surveillance and security systems. A report redacted by Mordor Intelligence [3] estimates that the video surveillance market has been valued at 30 billion dollars in 2016, but is expected to reach a value of 72 billion dollars by the end of 2022. A boost to the market perspective is also given by the recent results obtained in artificial intelligence and digital technologies—introducing intelligence, scalability, and higher accuracy in video surveillance solutions. Some spontaneous questions arise—what are the main technology trends in smart video surveillance and how can they be best used?

## 2. Technology trends in intelligent video surveillance solutions

- **Scene-aware intelligent video data gathering:** The obtained results in signal and image digital processing bring great progress to intelligent video surveillance systems, in particular, those that can be smarting adapt to the video data collection frame acquisition rate. When a security anomaly is detected, the data acquisition frame rate is increased accordingly in order to pick up richer and higher definition information for having more accurate and reliable results.

- **Big data infrastructures:** Advances in big data infrastructures have created more opportunities for video data storage and access, based on the four Vs of big data: volume, velocity, variety, and veracity. This way, gathering huge quantities of data from numerous cameras, taking into account high congestion streaming data rates, is much more efficient with respect to last year. Studying novel big data solutions increments the creation and deployment of smart video surveillance architectures that scale unceasingly and cost-effectively.

- **Streaming data devices:** Varying solutions in streaming systems have arisen in the last few years. These systems enable streaming management and analysis skills and are crucial portions of the big data systems examined.

- **Proactive analytics and artificial intelligence (AI):** Artificial intelligence and machine learning have given a new impulse for introducing new features in the smart surveillance systems, thanks to the materialization of disrupting deep learning methodologies, such as those introduced by Google's Alpha AI engine [4] and DeepMind. The growth of deep neural networks can be openly integrated into video surveillance systems to arm them with outstanding intelligence and able to boost more effective surveillance activities. As an interesting AI application includes predictive analytics, which helps security operators to foresee security events and act proactively.

- **UAV and the Internet of Things (IoT):** Next generation of smart surveillance systems and security includes the fusion of Internet of Things equipment and smart entities. Employing unmanned aerial vehicles (UAVs) (i.e., drones) in smart video surveillance introduces such versatility reaching some areas that are problematic to reach using traditional fixed cameras.

- **Physical and cyber security interaction:** Industrial assets' digital transformation with the innovation process is incrementally merging physical and digital security measures. New smart video surveillance solutions act as a protagonist in this merging because they express IT architectures employed to inspect wide physical areas. This way, they can be directly included in different cybersecurity structures for a universal and unified approach to security, safety, and surveillance.

## 3. Designing and developing of video surveillance systems

All the technologies described above open new challenges and possibilities in the expansion, application, and function of new generation of intelligent video surveillance systems. An important role is played by the developers and implementers of this intelligent surveillance systems, who should integrate and use full features of the mentioned cutting-edge technologies. To reach this objective, it is crucial to design and realize the right architecture for the video surveillance framework. Novel intelligent video surveillance solutions respect the **edge/fog-computing paradigm** [5] (see **Figure 1**) to elaborate video data sources earlier directly near the observed scene. Using this paradigm permits to save bandwidth performing real-time security supervising. Smart cameras are placed at the edge of the designed network and become edge nodes, where frames are grabbed and processed "in situ." This way, the intelligence is decentralized and these edge nodes can realize data collection intelligence and tuning frame rate, according to the recognized security context. Furthermore, they are linked to the cloud architecture, where information from multiple cameras is merged, assessed, and processed on higher time scales.

Choosing edge/fog-computing architectures [5] is the best choice for supporting the integration of past video surveillance systems with the actual technologies. IoT-driven drones will be combined with suitable edge nodes and they will be part of a mobile edge-computing infrastructure. It is strongly recommended that real-time processing of the acquired streaming flow should be computed at the edge, instead of in the cloud of the video surveillance architecture. Contrarily, deep learning computing can be performed both at the edge and in the cloud of the video surveillance infrastructure: If deep neural networks are placed at the edge, they can extract complex feature patterns in real time. However, the extraction of complex feature patterns and information over wider areas observed by several edge nodes (e.g., city-level structure) should be done only if deep learning is implemented in the cloud.

In general, it is difficult to find the right balance among the functionality to place in the cloud or on the edge. The decisions are done by making a trade-off among some opposite features (e.g., processing speed versus obtained results accuracy for some surveillance tasks).

All the smart video surveillance solutions should take some advantages of open equipment from different cameras and device vendors. In fact, a surveillance system may contain different devices and video capture means (e.g., high-definition cameras,

**Figure 1.**
*Edge/fog-computing paradigm.*

wired or wireless cameras, cameras in drones/UAVs, and so on). There are many benefit to have an open architecture because we can obtain flexibility, elasticity, and technological durability. In the last years, some energies have been spent to present an open, standards-based architecture for edge/fog computing in order to have video surveillance as the principal example of the use of fog computing.

## 4. Discussion and future works in intelligent video surveillance solutions

Choosing the right configuration when an edge-computing architecture is designed for implementing a video surveillance system that meets other kinds of challenges. These kinds of challenges include privacy preservation and data protection compliance. For example, producing surveillance devices is subjected to many privacy and data protection regulations and directives emanating from different countries. This often forces some restrictions in designing smart video surveillance systems. Stronger limitation are also applied in the use of drones that must respect some tighter regulations.

Another type of challenge interests the automation rate reached by the proposed solution. Considering that automation is commonly required for covering and monitoring wider spaces and saving further human workers, but human involvement in assessing is still necessary for the trustworthiness of the designed solution. Furthermore, it should be considered that nowadays new cyber-physical threats and attacks are arising against surveillance systems. Notice that a physical attack is often supported by a cyberattack on the video surveillance framework, which completely compromise the capacity to detect the physical assault that is happening.

The implementation of intelligence is data-driven (e.g., proactive threat prediction and AI analysis) needs large amounts of data that include examples of security threats that are very difficult to have. The study and design of artificial intelligence algorithms (e.g., lightweight and easy-to-use deep neural networks) is taking its first steps, although numerous innovative start-ups with cutting-edge AI products and services are already emerging.

Facing the many new challenges described above by developers and distributors of intelligent video surveillance solutions forces them to better comply with standards and regulations while adopting a phased approach to deployment. This gradual process should enable a transition from manual, that is, human-mediated, systems to fully automated video surveillance based on artificial intelligence.

Overcoming the challenges, we face requires a gradual implementation of data-driven intelligence. Starting with simple supervised training rules and moving on to more sophisticated machine learning techniques capable of detecting more complex asymmetric attack patterns. Another important outcome that could be achieved is the implementation of open architectures capable of accommodating innovative surveillance sensors by making them coexist with older ones, so as to exploit new advanced capabilities while obtaining the best value for money.

In conclusion, it can be said that all future smart video surveillance solutions may include many innovative features and functionalities, as they may employ new cutting-edge IT and artificial intelligence technologies.

## Author details

Pier Luigi Mazzeo
National Research Council of Italy (CNR), Institute of Applied Sciences and Intelligent Systems (ISASI), Lecce, Italy

*Address all correspondence to: pierluigi.mazzeo@cnr.it

## References

[1] Porikli F, Brémond F, Dockstader SL, Ferryman J, Hoogs A, Lovell BCS, et al. Video surveillance: Past, present, and now the future dsp forum. IEEE Signal Processing Magazine. 2013;**30**(3):190-198

[2] Xu Z, Hu C, Mei L. Video structured description technology based intelligence analysis of surveillance videos for public security applications. Multimedia Tools and Applications. 2016;**75**(19):12155-12172

[3] Available from: https://www.marketsandmarkets.com/Market-Reports/video-surveillance-market-645.html

[4] Available from: https://www.deepmind.com/research/highlighted-research/alphago

[5] Raj P, Saini K, Surianarayanan C, editors. Edge/Fog Computing Paradigm: The Concept, Platforms and Applications. Vol. 127. Advances in Computers Series; 2022. ISBN: 9780128245064

Chapter 2

# Change Point Detection-Based Video Analysis

*Ashwin Yadav, Kamal Jain, Akshay Pandey, Joydeep Majumdar and Rohit Sahay*

## Abstract

Surveillance cameras and sensors generate a large amount of data wherein there is scope for intelligent analysis of the video feed being received. The area is well researched but there are various challenges due to camera movement, jitter and noise. Change detection-based analysis of images is a fundamental step in the processing of the video feed, the challenge being determination of the exact point of change, enabling reduction in the time and effort in overall processing. It is a well-researched area; however, methodologies determining the exact point of change have not been explored fully. This area forms the focus of our current work. Most of the work till date in the area lies within the domain of applied methods to a pair or sequence of images. Our work focuses on application of change detection to a set of time-ordered images to identify the exact pair of bi-temporal images or video frames about the change point. We propose a metric to detect changes in time-ordered video frames in the form of rank-ordered threshold values using segmentation algorithms, subsequently determining the exact point of change. The results are applicable to general time-ordered set of images.

**Keywords:** change point detection, time-ordered images, difference image, threshold values, segmentation algorithms

## 1. Introduction

Intelligent video surveillance involves automated extraction of information related to an object or scene of interest, including detection, localization, and tracking amongst other applications. One of the earliest comprehensive efforts in this regard was undertaken by Robert T Collins et al. [1] as part of a Defence Advanced Project Research Agency (DARPA) Video Surveillance and Monitoring (VSAM) project. The three fundamental methods tested for moving object detection were the background subtraction, optical flow and temporal differencing method. Due to the limitation of individual methods, hybrid schemes involving combination of individual methods were tested. Adaptive background subtraction was combined with the three-frame difference method in order to overcome the limitation of either method. The frame difference method it may be noted is a simple technique but, however, suffers from the limitation that the complete shape of the detected object cannot be extracted precisely. The frame differencing method and in general temporal differencing make use of a static or dynamic threshold value in order to determine a change or no change scenario. This provides us a key to development of threshold as possible metric for our current work. Change detection (CD) is related to the fundamental task of object detection moving or static insofar as that it enables one to cull out relevant images or frames from a stack. Thus, the search space in scene analysis as part of the task for an image analyst gets reduced. This aspect is highlighted in the work by Huwer on adaptive CD for real-time surveillance applications [2]. CD enables one to detect viable changes, which can then be inputs for the subsequent object detection or tracking task. CD may be considered as an elementary stage in the video analytics framework entailing segmenting a video frame into the foreground and background. This may be considered a simple task but is an important precursor to further high-end processing. A most comprehensive recent review on a deep learning framework-based CD has been carried out by Murari Mandal et al. [3, 4]. Various applications of CD as part of video analysis including video synopsis generation, anomaly detection, traffic monitoring, action recognition, and visual surveillance have been covered as part of the study.

"*CD is the process of identifying differences in the state of an object or phenomenon by observing it at different times*" Singh [5, 6]. This standard definition of the CD process though applied to the context of remote sensing images articulates the objective and purpose clearly insofar as even video surveillance is concerned. The objective is to detect the relevant change as part of the video surveillance in form of the object or activity (phenomenon) of interest. Considering the fact that today the quantum of data in form of the video feed to be analysed by the image analyst has increased vastly in recent times, there is scope for automation in the analysis process at various levels. Determination of the exact change point (CP) within a set of video frames or sequences will reduce the workload of the image analyst by filtering in only the relevant changes that have occurred during the period of interest. This in turn shall increase the overall efficiency of the video analysis workflow by rendering the necessary automation as a useful aid to the analyst. Limited work in the domain of applied CD exists with regard to the aspect of determination of the exact point of change. This is the objective of the current work wherein we make use of the threshold of the difference image sequence based on various segmentation algorithms as a metric for the determination of the possible CP in an image sequence or video feed. Malek Al Nawashi et al. [7] have made use of the simple temporal differencing approach along with a threshold function in order to determine the moving image as part of their work on abnormal human activity detection in an intelligent video surveillance system. Thus, there is a scope to apply the image difference approach in order to determine the point of change while subsequently overcoming its limitation in terms of the inability to detect the complete target shape [1].

CP detection has been studied in time series data analysis. In the context of remote sensing images as a sample case from an image processing perspective, Militino et al. [8] have carried out a very comprehensive survey recently (2020), of the various methods and tools available for CP detection. They infer that the methods applied to time series data may be applied in the context of time-ordered satellite images and image processing as well. We would like to extend this notion to the case of image processing as applied to video analytics in general. Amongst the techniques studied the nonparametric approach is a viable option given the fact that abrupt changes are likely to occur in a video sequence at any point of time, rendering it difficult for an underlying Bayesian or model-based approach to be followed. The nonparametric approach is applicable to a wider variety of problems in CP detection since no assumption is made regarding any underlying model as surmised by Samaneh Aminikhanghahi et al. [9] in their comprehensive survey on CP detection methods for general time series data. The study points out that the inferences are applicable to the domain of image analysis as well. Nonparametric approach has also been analysed by Murari et al. [3, 4] too as part of their comprehensive survey on the DL-based CD methods as well.

One of the few studies on CP detection approach in a time-ordered set of images is that carried out by Manuel Bertoluzza et al. [10]. The objective of their work was to determine an accurate CD map between a selected pair of images amongst a time-ordered series of images by representing the changes along a temporal closed loop as binary sequences. In order to analyse the consistency of changes determined within a closed loop, the notion of a binary change variable was introduced. In our opinion, the use of the metric in order to compare the changes and finally achieve the desired accuracy is a novel idea. Though this step improves accuracy in existing methods of CD, the important question of determination of when a change has occurred or the CP still remains unanswered. The answer will enable efficient filtering of the video frames to a select few in form of image pairs about the respective CPs. The likely object or phenomenon of interest lies amongst these image pairs or frames. This can be a primary step yielding increased speed in processing within the overall intelligent video surveillance framework.

Based on the above discussion, the objective of our study is to determine a simple and robust method to determine CP within a set of segmented video frames forming part of a video surveillance feed. A change of variable or metric [10] in form of the threshold based on different image pairs is utilized to determine the point of change from amongst a set of images or frames. Rank ordering the changes based on the thresholds enables second or third CP detection as deemed fit by the image analyst. Nonparametric methods are more robust making no assumptions about the underlying model structure [9] and amongst these, Pettitt's approach [11] is a simple and widely used technique. Our proposal for the change metric is similar to that of Pettitt's.

The main contribution of the work is the following: 1) Determination of a suitable CD metric based on a comparative analysis of various segmentation methods to include Otsu, K Means based (denoted by ISODATA), minimum cross-entropy threshold (MCE) methods, 2) Application of the CD metric CP detection within the set of segmented video frames, and 3) Proposed framework to apply the CD metric-based CP concept to the intelligent video surveillance problem.

The chapter is organized as follows. Subsection 1.1 after introduction covers the aspects of the data set. Section 2 describes the basic CP detection algorithm based on the change metric concept. Section 3 discusses the results obtained based on a comparative analysis of respective CD metrics obtained from the four segmentation algorithms tested. Section 4 describes the proposed application framework of the results to the intelligent video surveillance framework.

### 1.1 Data set description

Most CD open source data sets are in form of image pairs as the objective is the application and testing of specific CD methods or algorithms to the same. In order to achieve the objective of the current work, there is a need for a time-ordered image data set. For this purpose, Google Earth-based time-ordered satellite image data sets of specific locations sourced from open source data [12] have been used and customized for testing purposes. The satellite image data set has viability for automation in terms of information extraction by the image analyst, which is currently being done manually. Hence, the choice of this data set for developing results as part of the study has been undertaken. However, it is worth mentioning that the results obtained subsequently can be well applied to a general image processing scenario including video analysis. Google Earth images are a valid source of satellite imagery used for research purposes as evinced in work such as Ur¨ka Kanjir's et al.'s survey [13]. The sample data set is as shown in **Figures 1-3**.Out of the time-ordered data set of 19 images, the relevant point of change is that between the fourth and fifth image (refer red arrow in **Figure 1**) when the object of interest or change has first appeared. The testing has been carried out on 10 such sets with the object appearing at an instance within the data set, which denotes the point of change. The spatial resolution of the data set is as per the standard Google Earth platform (= > 5 cm) with each image corresponding to an area of 12 x 12 km on ground. The average temporal resolution of the 10 data sets of images was 10–15 years calculated between the first and last set of images.

CDNET2014 [14] is another standard open source data set for testing various CD algorithms based on static images and video sequences. We make use of this data set to demonstrate a more general application of the algorithm and analyse results on a test case along with those obtained for the above cases (**Figures 1–3**). The data set sample pertains to the intermittent object motion category and is like a parking lot with a man entering the scene at a certain point (frame number 57). **Figures 4** and **5** show the sample data set that actually consists of 2500 frames forming part of a video feed in which testing is carried out on a selected number of frames (e.g. 80). The objective is to detect the point of change which is at the point of entry of the

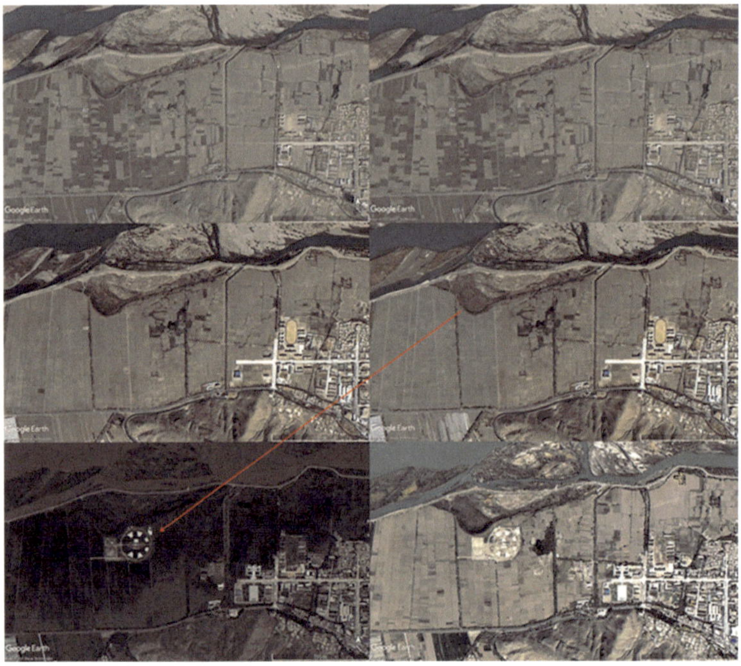

**Figure 1.**
*Sample data set sequence 1.*

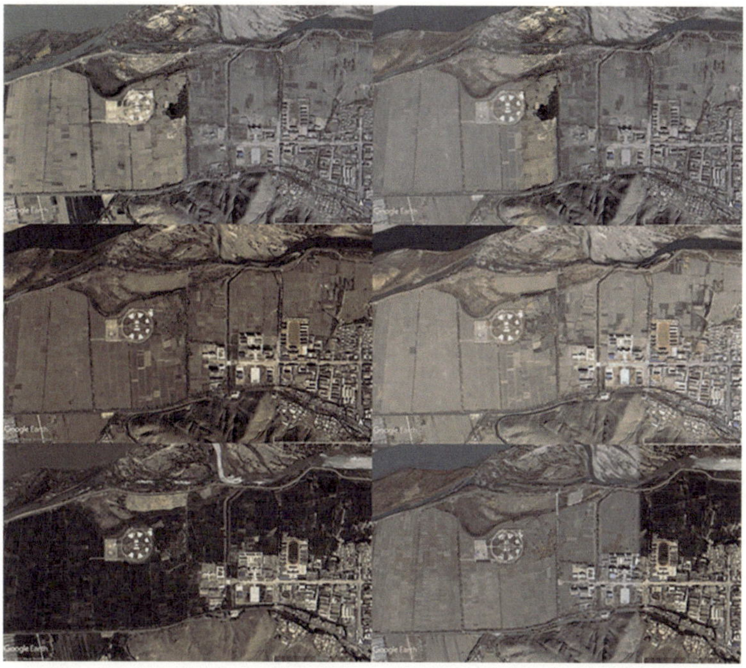

**Figure 2.**
*Sample data set sequence 2.*

individual. As can be observed, the changes are extremely minor and difficult to detect between respective frames as it is of a video recording. Application of various segmentation algorithms such as Otsu, MCE, ISODATA and analysis therein to the Google Earth and CDNET2014 data set shall enable the selection of a suitable method accordingly.

## 2. Concept: Change point detection

### 2.1 Background

CP detection in time series is a well-researched area with a comprehensive survey on various methods carried out by Aminikhanghahi et al. [9]. The application areas include medical condition monitoring, climate change monitoring, speech recognition and image analysis. CP detection in image analysis is the least researched area, and our endeavour in the current work is to apply the useful lessons learned in the case of the time series approach to that of image or video analysis. CP detection in time series is much simpler compared to the case of image or video analysis considering that the numeric values to be compared are easily extracted from the data itself. CP detection in case of image or video analysis requires the determination of a suitable change metric to be applied in a similar framework of time series in order to apply the benefits of the same in this case. Trend and CP detection in remote sensing has been well studied and classified by Militino et al. [7]. The nonparametric methods are robust and applicable to a larger variety of problems compared with parametric

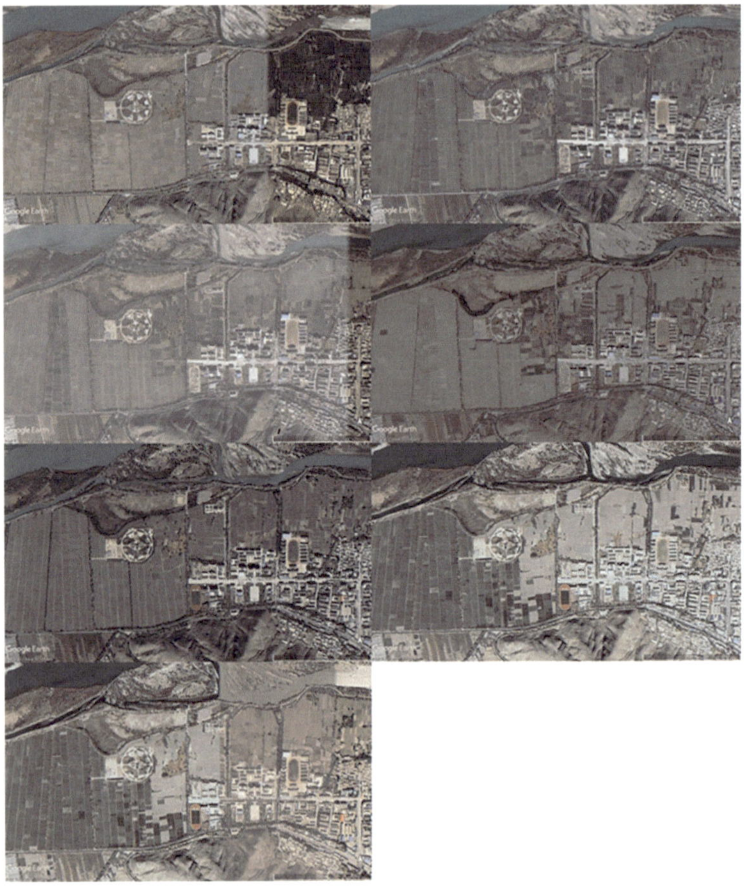

**Figure 3.**
*Sample data set sequence 3.*

methods since changes in phenomenon or objects may be arbitrary not following any pattern or model. Amongst nonparametric methods, Pettitt's method [11] is a well-established and applied method. We take a cue from this approach wherein the random variables forming part of the test hypothesis are substituted by the respective threshold values of the difference image sequences in order to determine the CP as explained below.

A suitable change variable or metric for the determination of the maximum CP in a time-ordered image set is the threshold values obtained from image segmentation of the different pairs of images. Subject to a minimum or no change scenario between images there will be minimum or no variation amongst the respective threshold values in the set. This premise has a rationale that any change in the sequence of images shall result in a variation in pixel values. This variation can be directly captured in form of a variation in the threshold values of the segmented image as per different algorithms applied. Otsu Binary segmentation algorithm [15] is a standard segmentation algorithm along with Li's information theoretic MCE threshold method [16] and Coleman's K means clustering image segmentation algorithm [17]. The threshold

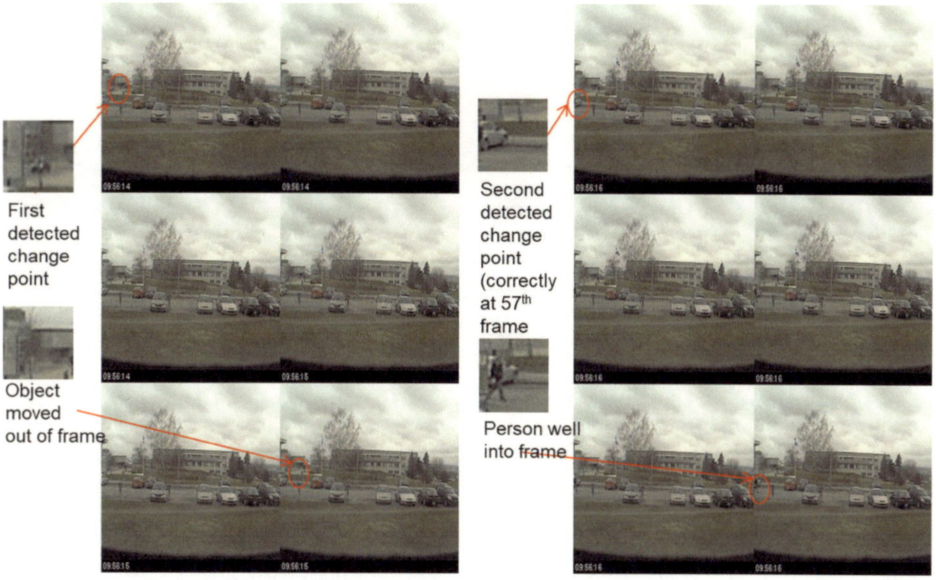


**Figure 4.**
*CDNET2014 data set result (MCE).*

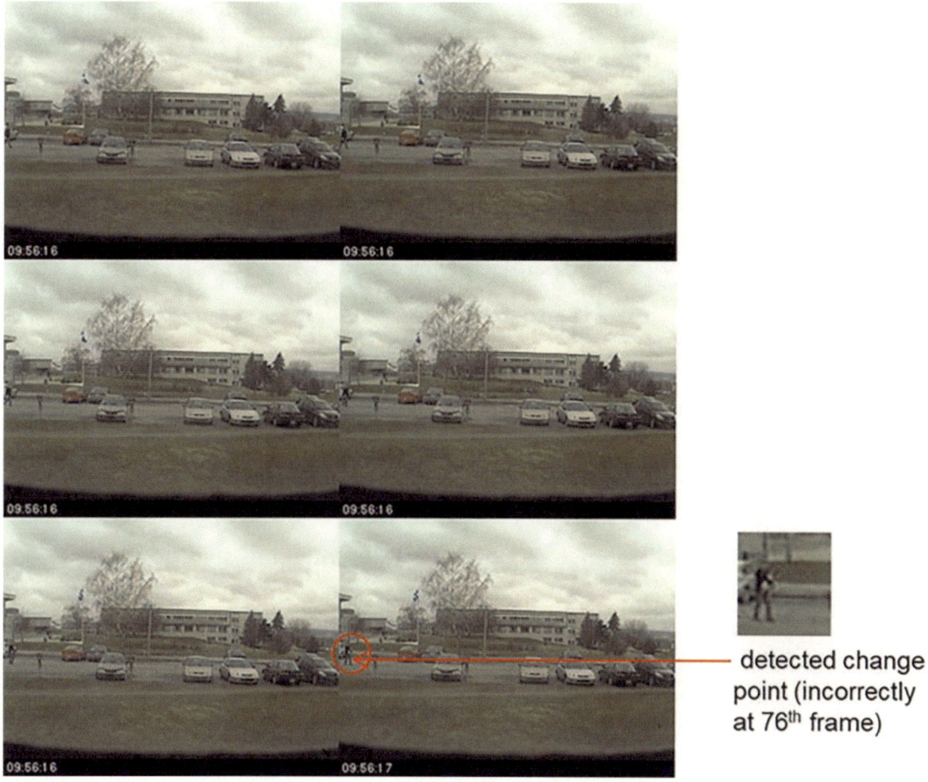


**Figure 5.**
*CDNET2014 data set result (Otsu).*

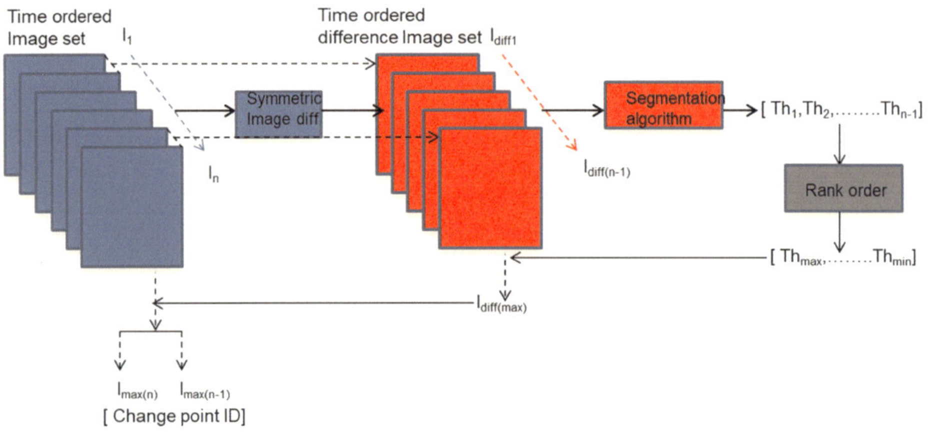


**Figure 6.**
*Proposed basic CD framework.*

values determined by these algorithms along with a mean threshold method are proposed to be used as the change variable or metric for the determination of CP in the time-ordered image sequence.

The methodology is thus based on thresholding (*via* application of the respective segmentation algorithms) of the binary image difference sequences constituting the image set. The point of maximum change is determined by the maximum value from amongst the threshold sequences of the binary image difference sequence. The algorithm is described in steps in the next section as illustrated in **Figure 6**.

## 2.2 Steps

Let us consider the set of time-ordered image sequences or video frames as $T = \{I_n|n\varepsilon[1,\ N]\}$ where $N$ is the total number of the images being processed. The objective is to select the pair of images that define precisely the maximum CP and further rank order the images in reducing relevance of CPs. This will assist the image analyst in sifting the images so as to determine the exact point of change while analysing the phenomenon or object of interest. This will enable timely and efficient analysis of the time-ordered image sequences or video frames. The steps are as follows:

1. Determine the image difference sequence (e.g. based on the symmetric difference *absdiff* method in python) as $T_{diff} = \{I_1 - I_2, I_2 - I_3, \ldots \ldots, I_{n-1} - I_n\}$.

2. Segment the image difference sequence based on methods such as [15–17] $S(T_{diff}) = S[\{I_1 - I_2,\ I_2 - I_3,\ \ldots \ldots,\ I_{n-1} - I_n\}]$ and buffer the respective threshold values as $T_{th} = \{Th_1, Th_2, Th_3, \ldots \ldots, Th_{n-1}\}$, $T_{th} = \{Th_n|n\varepsilon[1,\ N-1]\}$.

3. On lines of Pettitt's method [11] the CP in terms of the threshold is determined as $CP = max\,[T_{th}]$.

4. Rank order the sequence of threshold values from maximum to minimum to determine CPs in decreasing order of relevance to aid the image analyst.

5. Based on the index of CP the corresponding image pair may be processed further to extract information as desired by the image analyst.

## 3. Results and analysis

### 3.1 Results

Methodology and steps described in section 2 have been applied to 10 data sets of the type described in **Figures 1–3**, and results obtained therein are displayed in **Tables 1** and 2, respectively. **Table 1** pertains to the category 1 evaluation wherein no margin for error is permitted and a valid detection is considered if as per ground truth, the CP is detected based on the maximum threshold value of the segmented difference image sequence. This is in keeping with the requirements or validity of the algorithm. It is also possible that due to pixel value variations owing to noise, and in certain cases the precise point of change is not captured corresponding to the maximum threshold value but the second highest threshold value or subsequent. Corresponding to this relaxation (valid detection considered up to the second highest threshold value), the results are re-valuated and presented in **Table 2** as category 2. The standard Receiver Operator Characteristic (ROC) metrics of True Positive (TP), True Negative (TN), False Positive (FP) and False Negative (FN) are applicable for the current methodology as well with a slight modification. A correct detection in form of a TP corresponds to a TN as well since we are interested in the detection of only the correct image pair and not in the number of targets detected in a particular image as per standard applications. Similarly, in case, the correct image pair is not detected, that is, a FP occurs that corresponds to a FN as well. Recalling as per the standard

| Method | TP | TN | FP | FN | Recall (%) |
|---|---|---|---|---|---|
| Otsu | 8 | 8 | 2 | 2 | 80 |
| MCE | 9 | 9 | 1 | 1 | 90 |
| ISODATA | 8 | 8 | 2 | 2 | 80 |
| Mean | 6 | 6 | 4 | 4 | 60 |

**Table 1.**
*ROC metrics: category 1.*

| Method | TP | TN | FP | FN | Recall (%) |
|---|---|---|---|---|---|
| Otsu | 10 | 10 | 0 | 0 | 100 |
| MCE | 10 | 10 | 0 | 0 | 100 |
| ISODATA | 10 | 10 | 0 | 0 | 100 |
| Mean | 7 | 7 | 3 | 3 | 70 |

**Table 2.**
*ROC metrics: category 2.*

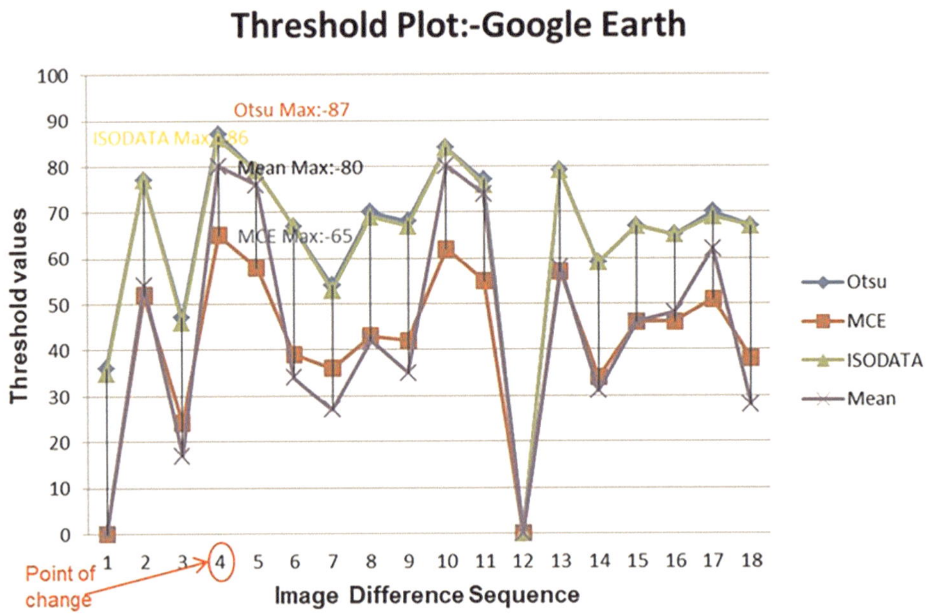


**Figure 7.**
*Threshold plot: Google Earth sample data set.*

definition = (TP/(TP + FN)) represents the number of valid targets correctly detected. Precision as per the standard definition = (TP/(TP + FP)) gives the quality of the detection in terms of the correct number of target detected with a minimum of FPs. F1 Score as per the standard definition = (2*Precision*Recall)/(Precision+Recall) represents the degree of balance obtained in terms of both the precision and recall. In the present case for reasons aforesaid, that is, FP and TP being coincidental with FN and TN, respectively, the values of Recall, Precision and F1 score will all give the same values. Hence, for ease of assimilation of the reader and analysis therein we only mention Recall. Certain applications such as military target detection call for a high degree of recall compared with precision, that is, a minimum or no target miss scenario wherein one is ready to compromise to a certain extent on precision viz.-a-viz. recall.

The threshold plot corresponding to the sample image sequence (refer **Figures 1–3**) is presented in **Figure** 7. The threshold plot displays the variation in values corresponding to the respective threshold methods. As is visible, the point of change is correctly detected between the fourth and fifth image, compared with the ground truth (refer red arrow in **Figure 1**).

The CDNET2014 data set sample results are shown in **Figures 4** and **5**, respectively, with the corresponding plot displayed in **Figure 8**. From **Figure 4**, it is observed that using the cross-entropy method, MCE is able to detect the point of change accurately (refer **Figure 4**) with Otsu method (refer **Figure 5**).

### 3.2 Analysis of results

Based on the results, following are the relevant deductions:

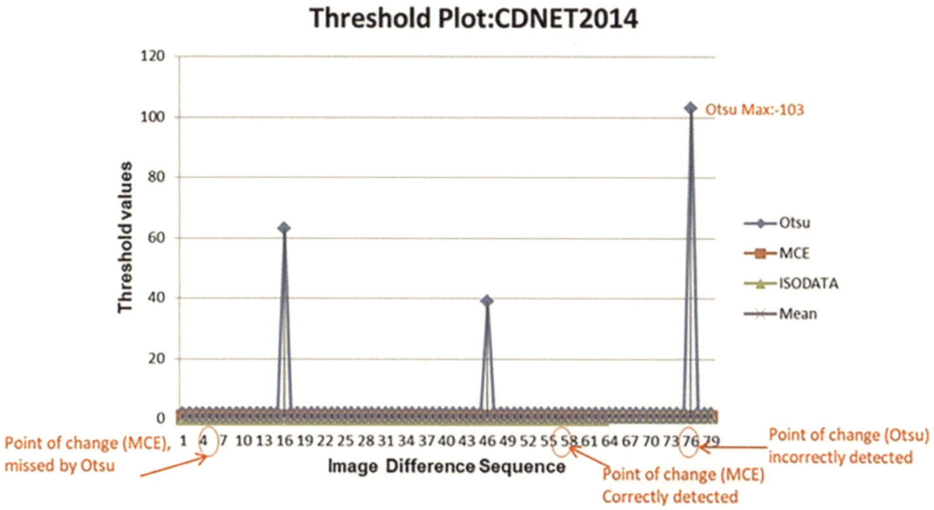


**Figure 8.**
*Threshold plot: CDNET2014 data set.*

1. From **Tables 1** and **2**, and plot in **Figure 7**, it is observed that the three methods Otsu, MCE and ISODATA perform well and are able to detect the CP accurately in case of the Google Earth data set.

2. For the category 1, the metric value of MCE has a slight edge compared with the other two methods, that is, Otsu and ISODATA as seen in **Table 1**. This is important considering that it is an information theoretic approach. As per the threshold plot in respect of MCE, it is observed that a greater capability to distinguish CPs exists.

3. The plot of **Figure 8** along with **Figures 4** and **5** provides another dimension to compare the methods based on the standard CDNET2014 data set [14]. It is observed that when there is a minute variation in changes between images in a video frame format, MCE is the only method that can distinguish the changes and accurately determine the relevant point of change. This is due to the fact that when a minor target enters a frame, the Otsu method tends to shift the threshold towards the foreground thereby suppressing relevant details [18]. Similarly, the ISODATA and mean methods also do not yield the correct results. The entry of the target (person) into the frame is incorrectly detected by the Otsu method as bit late in 76th frame (refer **Figure 5**) as compared with the actual frame in which the person enters, that is, 57th detected correctly by MCE (refer **Figure 4**). The CDNET2014 data set results to corroborate the findings given in **Table 1** wherein the cross-entropy method provides the best performance.

4. The segmentation methods in order of performance are ranked as MCE followed by Otsu, ISODATA and lastly mean method. The cross-entropy has a slight edge over the Otsu, which has been observed in the Google Earth data set case (refer **Table 1**) while being validated on the CDNET2014 data set.

5. Irrespective of the CD method used for example, image difference-based approach or transformation-based approaches such as Principal component analysis (PCA), it is observed that the change metric in form of the threshold values of the segmentation method is a viable option for detecting point of change as validated based on the two data sets described above.

6. The results thus obtained can be well applied to a general image processing scenario including the application towards intelligent video surveillance.

## 4. Proposed framework: CP detection in video analysis

### 4.1 Case I: Static format

Based on the basic CD concept described in section 2 and results obtained in section 3, we describe two formats for implementation as part of the intelligent video analysis framework. The current description in this subsection pertains to static format case (refer Case I in **Figure 9**) wherein only a limited number of video frames or images are received and required to be analysed. In this scenario, the determination of the important CPs and in turn the filtering of probable objects or phenomenon are based on the basic CD framework described in Section 2. The steps remain the same as described in subsection B of the section. **Figure 1** is a diagrammatic description of the concept, which is further modified for video surveillance case vide as in **Figure 9**. The modification is the addition of level I or level 2 processing element in form of a basic segmentation algorithm or an object detection algorithm. The level 1 processing scenario entails application of a segmentation algorithm as used for change metric determination, applied to the different images or image pairs about the CP. In case of searching for a specific category of target, a level 2 processing step in form of an object detection algorithm may be applied. A level 1 processing step applies the same segmentation algorithm (e.g. MCE) that has been used to determine the CD metric. This

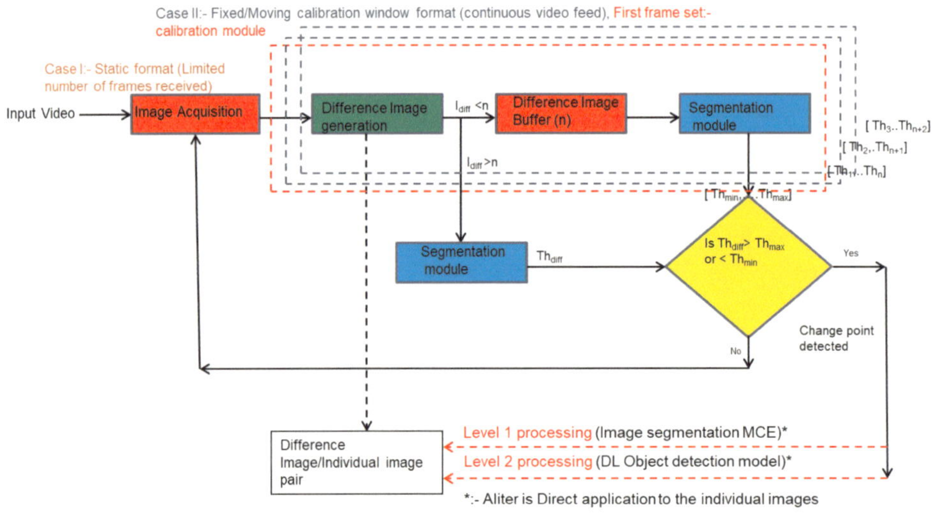


**Figure 9.**
*Proposed CD framework for video analytics.*

ensures full exploitation of the notion of a segmentation algorithm in terms of its capability to distinguish or partition a scene into the foreground and background [3, 4]. The foreground is likely to contain the phenomenon of interest or object of interest. By filtering the entire set of images or video frames received, to a likely pair of images the overall time period of processing will definitely reduce and effort too on the part of the image analyst. The steps as described vide in subsection B in Section 2 are applicable in the current case too and are as follows:

1. Determine the image difference sequence (e.g. based on the symmetric difference *absdiff* method in python) as $T_{diff} = \{I_1 - I_2, I_2 - I_3, \ldots \ldots, I_{n-1} - I_n\}$.

2. Segment the image difference sequence based on methods such as [15–17] $S(T_{diff}) = S[\{I_1 - I_2,\ I_2 - I_3,\ \ldots \ldots,\ I_{n-1} - I_n\}]$ and buffer the respective threshold values as $T_{th} = \{Th_1, Th_2, Th_3, \ldots \ldots, Th_{n-1}\}$, $T_{th} = \{Th_n | n\varepsilon[1,\ N-1]\}$.

3. On lines of Pettitt's method [11] the CP in terms of the threshold is determined as $CP = max\,[T_{th}]$.

4. Rank order the sequence of threshold values from maximum to minimum in order to determine CPs in decreasing order of relevance in order to aid the image analyst.

5. Based on the index of CP, the corresponding image pair may be processed further to extract information as desired by the image analyst.

### 4.2 Case II: Fixed/moving calibration window format

This format is applicable when a continuous feed of video frames is being received for analysis in a fixed or moving camera scenario. The fixed window implies application of a calibration module over the first set of frames (refer red box and title First frame set). As part of the calibration module, the corresponding thresholds of the difference image sequences are determined. Once all calibration frames are received, the minimum and maximum thresholds corresponding to the segmented difference images are determined. The premise of employing a calibration module is to capture the background model in form of the thresholds of the successive difference images prior to the system being applied in a live scenario. Live scenario pertains to the actual phase of application wherein the information regarding the object or scene of interest is to be captured. Thus, in order to analyse the environment or background where the fixed or moving camera is employed, the calibration module enables capturing the background information or no change detected scenario. Once the thresholds of successive difference images are captured as part of the calibration module, subsequent threshold of difference images lying outside the range of those of the calibration module is indicative of a probable CD scenario. The yellow rhombus indicates this decision in **Figure 9**.The issues of false triggering are likely to be reduced as minor variations in the scene, which are to constitute the background captured in the calibration module prior to the application of the live phase. The steps for fixed calibration window are as follows:

1. Determine the image difference sequence for the say first $n$ image difference sequence forming part of the calibration frame set as $T_{diff} = \{I_1 - I_2, I_2 - I_3, \ldots \ldots, I_{n-1} - I_n\}$.

2. Segment the image difference sequence based on methods such as [15–17] $S(T_{diff}) = S[\{I_1 - I_2,\ I_2 - I_3,\ \ldots \ldots,\ I_{n-1} - I_n\}]$ and buffer the respective threshold values as $T_{th} = \{Th_1, Th_2, Th_3, \ldots \ldots, Th_{n-1}\}$, $T_{th} = \{Th_n | n\varepsilon[1,\ N-1]\}$.

3. Bracket the max and minimum thresholds as $CP = [max\ [T_{th}], [min\ [T_{th}]]$.

4. From image sequence number $n+1$ and beyond post-application of the calibration module described in above steps, the live phase commences wherein each image difference pair starting with $I_{n+1} - I_n$ is tested for being a valid CP by segmenting the same to yield thresholds starting with $Th_{n+1}$. Thus, the step is Check if $min\ [T_{th}] < Th_{n+1} < max\ [T_{th}]$, that is, whether $Th_{n+1}$ lies in CP. If no, then the image pair constitutes a valid CP. This step is represented by the yellow rhombus in **Figure 9**.

5. If the current difference image constitutes a CP, then apply the level 1 or level 2 processing for further analysis or else repeat step 5.

6. Based on the application of the level 1 or level 2 processing present results with regard to the probable target or scene of interest duly processed as an aid to the image analyst.

The moving window concept is similar to the static case with the difference that the corresponding maximum and minimum threshold values vary as per the shifting window or set of frames over which calibration is carried out. In this case, the problem of dynamically changing scenarios such as vehicles starting and stopping abruptly is addressed. In such cases, the background needs to be dynamically updated for which adaptive algorithms have been proposed [1]. However, the CD metric is a powerful concept which in the current scenario is representative of the background static or dynamic as captured in form of the calibration module. In case of an envisaged scenario wherein the dynamic variation in background continues for a longer period, the moving window calibration module is applied to overcome these problems. Here, the threshold ranges detected over a fixed calibration frame within the static format are varied to change over sequences of frames being captured. The moving window calibration frames are depicted *via* the dashed lines in **Figure 9**. As the video frames are received, the set of thresholds corresponding to the calibration module are captured over the latest set of video frames in a pre-decided interval (corresponding to the anticipated degree of dynamism in background). Thus, the range of threshold values of the calibration module will be shifted over the next set of say $n$ video frames thereby capturing the latest background in order to detect corresponding changes in subsequent frames. The steps for moving calibration window are as follows:

1. Determine the image difference sequence for the say first $n$ image difference sequence forming part of the calibration frame set as $T_{diff} = \{I_1 - I_2, I_2 - I_3, \ldots \ldots, I_{n-1} - I_n\}$.

2. Segment the image difference sequence based on methods such as [15–17] $S(T_{diff}) = S[\{I_1 - I_2,\ I_2 - I_3,\ \ldots\ \ldots,\ I_{n-1} - I_n\}]$ and buffer the respective threshold values as $T_{th} = \{Th_1, Th_2, Th_3, \ldots\ \ldots, Th_{n-1}\}$, $T_{th} = \{Th_n | n\varepsilon[1,\ N-1]\}$.

3. Bracket the max and minimum thresholds as $CP = [max\,[T_{th}],[min\,[T_{th}]]$.

4. From image sequence number $n+1$ and beyond post-application of the calibration module described in above steps the live phase commences wherein each image difference pair starting with $I_{n+1} - I_n$ is tested for being a valid CP by segmenting the same to yield thresholds starting with $Th_{n+1}$. Thus, the step is Check if $min\,[T_{th}] < Th_{n+1} < max\,[T_{th}]$, that is, whether $Th_{n+1}$ lies in CP. If no then the image pair constitutes a valid CP.

5. If the current difference image constitutes a CP, then apply the level 1 or level 2 processing for further analysis or else repeat step 5.

6. Based on the application of the level 1 or level 2 processing present results with regard to the probable target or scene of interest duly processed as an aid to the image analyst.

7. The steps from 1 to 6 are repeated by modifying the calibration frame set starting with step 1 as $T_{diff} = \{I_{t+1} - I_{t+2}, \ldots\ \ldots, I_{t+n-1} - I_{t+n}\}$. It may be noted that $t$ pre-decided number of frames after which recalibration is carried out in terms of the fresh set of frames. The setting of the value of $t$ corresponds to the degree of dynamism anticipated in terms of the changing background wherein erstwhile foreground elements are anticipated to merge with the background. Thus, the least value of $t$ set to 1 corresponds to a highly dynamic scenario wherein the foreground elements tend to merge with the background rapidly.

The limitation in the simple frame differencing method of being unable to recover a complete shape of detected target [1] is overcome in our proposed framework by application of a Level 1 or Level 2 processing step post-detection of the CP as shown in **Figure 9**. Thus, once the point of change is detected, further application of say a level 2 processing will enable determination of the complete shape of the intended target.

### 4.3 Implementation issues

The CP detection-based methodology proposed for video analysis as described above in subsections A and B, respectively, is a simple adaptation of the CP-based approach. The advantage of the approach as adapted for video analysis is that it is simple and independent of the time-ordered set of video frames being received. Both offline (refer subsection A) and online (refer subsection B) options of implementation exist and it is a nonparametric approach, not making any assumptions regarding the underlying model. The change metric is a single value derived in a simple manner independent of any probabilistic methodology. Thus, the approach being nonparametric is applicable to a large number of scenarios since no assumptions are made regarding any specific scenario. The methodology is unsupervised not requiring any training data as in case of many deep learning or machine learning-based approaches.

Thus, the speed of implementation will be inherently higher in our case. The challenges in application of the method proposed are that initially it will require certain amount of testing and fine tuning in conjunction with an image analyst (for checking the performance of the algorithm). Factors such as the number of calibration frames, that is, window size for determination of the CD metric, will require certain fine tuning and innovation during implementation stage. The basic CP framework as described in Sections 2 and 3 was executed in Python code and the adaptation for the video analysis framework as described in the current section may follow suite. The architecture described in **Figure 9** is simple and flexible and may hence be modified suitably as per results obtained during implementation stage.

### 4.4 Comparison with the state of the art (SOA) in intelligent video surveillance

The current focus of the SOA in the field of video surveillance is primarily on specific application scenarios as described in the comprehensive review by Guruh Fajar Shidik et al. [19]. Intelligent video surveillance includes anomaly detection, object detection, target tracking, etc., as few of the applications, which could apply a CD algorithm component as an important precursor step. It is worth noting that the CP detection concept as described in Sections 2 and 3 covering the application to the video analysis framework has not been well researched. Hence, a valid comparison with an equivalent method in context of video analytics does not exist. The closest semblance to the proposed method based on the CD concept is that of a discriminative framework for anomaly detection proposed by Allison Del Giorno et al. [20]. The proposed method endeavours to overcome the limitation in the existing anomaly detection methods namely the requirement of training data and dependence on temporal ordering of the video data. Their method is based on a nonparametric technique inspired by the density ratio estimation for CP detection. The approach is novel and similar to our proposed method in terms of the nonparametric approach wherein no assumptions are made about the underlying model. Further, the method proposed by Allison et al. does not require training data and is unsupervised similar to our case as well. They endeavour to use a metric- or score-based approach in order to determine anomaly points in a video sequence independent of the ordering of the video frames. However, the method does require an input of the features to aid in distinguishing the anomalies. It may be noted that the proposed methodology in our case is much simpler wherein no such feature set description is required to determine the CP and a single metric in form of the threshold of the image difference pairs is sufficient. This metric-based approach in our case makes the method simple and fast. Moreover, the CP concept is robust and adaptable to an anomaly detection framework. Thus, our method is simpler than the approach proposed by Allison Del Giorno et al. [20], which ultimately utilizes a probabilistic approach to determine the metric used to determine the anomaly points. The proposed CP-based video analysis methodology may be considered as a primary step in the intelligent video analysis framework prior to application of subsequent steps and a potential field for research. This analysis is to the best of the knowledge of the authors, the most relevant possible comparison with the SOA. A thorough review of the existing CD methods in other areas such as time series analysis and remote sensing has already been covered as part of literature review in Section 1. Thus, Section 1 and the current subsection comprehensively cover all the aspects of the proposed method and its typesetting viz.-a-viz. other areas of research.

## 5. Conclusion

To the best of the knowledge of the authors, this is the only study on CP detection in respect of image processing in particular as applicable to video surveillance as well. Important results have been obtained with the best method being determined as the cross-entropy MCE, followed by Otsu and ISODATA thereafter. The image difference-based CD metric method is by no means limited only to time-ordered set of images as represented in **Figures 1–3**. The method has been applied to a selected CDNET2014 data set as well as displayed in **Figures 4** and **5**. It may be noted that the sequence of images taken from the CDNET2014 data set are originally part of a video sequence, and hence, the results demonstrated in Section 3 (Refer **Figures 4** and **5**) are well suited to be applied to a video surveillance scenario. Thus, formulating the method in sliding window format will enable application to video surveillance scenario including suspicious activity detection scenarios. The block diagram for the proposed application of the CD concept is displayed in **Figure 9** and proposed methodology has been described in detail in Section 4. The scope of applications possible is by no means limited to these two cases. In summary, the CD metric methodology in form of the threshold value needs to be exploited in an innovative manner. Further alternate change variable metrics may be a good area for further research. The objective of the current work has been to answer the important question of Where the change lies? or when has it occurred in a time-ordered set of images? This is important in order to act as a precursor for pin-pointed analysis of the images about the detected point of change as proposed in Section 4.

The level 2 processing in **Figure 9** may also be in an Object Based CD (OBCD) framework [21]. Alternate options for processing the images detected about the CP may be considered part of future research scope.

## Author details

Ashwin Yadav[1*], Kamal Jain[1], Akshay Pandey[1], Joydeep Majumdar[2] and Rohit Sahay[3]

1 Department of Civil Engineering, Indian Institute of Technology, Roorkee, India

2 Department of Mechanical Engineering, Indian Institute of Technology, Indore, India

3 Department of Computer Science Engineering, Indian Institute of Technology, Kharagpur, India

*Address all correspondence to: ashwiny77@gmail.com

## References

[1] Collins R, Lipton A, Kanade T, Fujiyoshi H, Duggins D, Tsin Y, et al. A System for Video Surveillance and Monitoring Tech. Report, CMU-RI-TR-00-12, Robotics Institute, Carnegie Mellon University; May 2000

[2] Huwer S, Niemann H. Adaptive change detection for real-time surveillance applications. Proceedings Third IEEE International Workshop on Visual Surveillance. July 2000. pp. 37-46. DOI:10.1109/VS.2000.856856

[3] Mandal M, Vipparthi SK. An Empirical Review of Deep Learning Frameworks for Change Detection: Model Design, Experimental Frameworks, Challenges and Research Needs, in IEEE Transactions on Intelligent Transportation Systems. July 2022;**23**(7):6101-6122. DOI: 10.1109/TITS.2021.3077883

[4] Lu D, Mausel P, Brondízio E, Moran E. Change detection techniques, International Journal of Remote Sensing. 2004;**25**(12):2365-2401. DOI: 10.1080/0143116031000139863

[5] Singh A. Review Article Digital change detection techniques using remotely-sensed data. International Journal of Remote Sensing. 1989;**10**(6): 989-1003. DOI: 10.1080/01431168908903939

[6] Isever M, Ünsalan C. Two-Dimensional Change Detection Methods: Remote Sensing Applications. Springer Publishing Company, Incorporated; 2012. ISBN: 978-1-4471-4254-6

[7] Al-Nawashi M, Al-Hazaimeh OM, Saraee M. A novel framework for intelligent surveillance system based on abnormal human activity detection in academic environments. Neural Computation and Application. 2017; **28**(1):565-572. DOI: 10.1007/s00521-016-2363-z

[8] Militino AF, Moradi M, Ugarte MD. On the performances of trend and change-point detection methods for remote sensing data. Remote Sensing. 2020;**12**(6):1008. DOI: 10.3390/rs12061008

[9] Aminikhanghahi S, Cook DJ. A survey of methods for time series change-point detection. Knowledge and Information Systems. May 2017;**51**(2): 339-367. DOI: 10.1007/s10115-016-0987-z. Epub 2016 Sep 8. PMID: 28603327. PMCID: PMC5464762

[10] Bertoluzza M, Bruzzone L, Bovolo F. A novel framework for bi-temporal change detection in image time series. 2017 IEEE International Geoscience and Remote Sensing Symposium (IGARSS), 2017. pp.1087-1090. DOI: 10.1109/IGARSS.2017.8127145

[11] Pettitt AN. A non-parametric approach to the change-point problem. Journal of the Royal Statistical Society: Series C (Applied Statistics). 1979;**28**(2): 126-135. DOI: 10.2307/2346729

[12] Available from: http://climateviewer.org/history-and-science/government/maps/surface-to-air-missile-sites-worldwide

[13] Kanjir U, Greidanus H, Oštir K. Vessel detection and classification from spaceborne optical images: A literature survey. Remote Sensing of Environment. 2018;**207**:1-26. ISSN 0034-4257. DOI: 10.1016/j.rse.2017.12.033

[14] Wang Y, Jodoin P-M, Porikli F, Konrad J, Benezeth Y, Ishwar P. CDnet 2014: An Expanded Change Detection Benchmark Dataset. United States: IEEE CVPR Change Detection workshop. Jun 2014. p. 8. (hal-01018757)

[15] Otsu N. A threshold selection method from gray-level histograms. IEEE Transactions on Systems, Man, and Cybernetics. Jan 1979;**9**(1): 62-66. DOI: 10.1109/TSMC.1979. 4310076

[16] Li CH, Lee CK. Minimum cross entropy threshold. Pattern Recognition. 1993;**26**(4):617-625. DOI: 10.1016/0031-3203(93)90115-D. ISSN 0031-3203

[17] Coleman GB, Andrews HC. Image segmentation by clustering. Proceedings of the IEEE. 1979;**67**(5):773-785. DOI: 10.1109/PROC.1979.11327

[18] Malik MM, Spurek P, Tabor J. Cross-entropy based image thresholding. Schedae Informaticae. 2015;**24**:21-29. DOI: 10.4467/20838476SI.15.002.3024

[19] Shidik GF, Noersasongko E, Nugraha A, Andono PN, Jumanto J, Kusuma EJ. A systematic review of intelligence video surveillance: Trends, techniques, frameworks, and datasets. in IEEE Access. 2019;7:170457-170473. DOI: 10.1109/ACCESS. 2019.2955387

[20] Giorno AD, Bagnell JA, Hebert M. A discriminative framework for anomaly detection in large videos. In: Leibe B, Matas J, Sebe N, Welling M, editors. Computer Vision – ECCV 2016. ECCV 2016. Lecture Notes in Computer Science. Vol. 9909. Springer, Cham. 2016. DOI: 10.1007/978-3-319-46454-1_21

[21] Hussain M, Chen D, Cheng A, Wei H. David Stanley, change detection from remotely sensed images: From pixel-based to object-based approaches. ISPRS Journal of Photogrammetry and Remote Sensing. 2013;**80**:91-106. ISSN 0924-2716

Chapter 3

# Spaceborne Video Synthetic Aperture Radar (SAR): A New Microwave Remote Sensing Mode

*Jian Liang and Liang An*


## Abstract

The transient information like a 'picture' can be obtained by the traditional microwave remote sensing system. It will bring some shortcomings for detection of the moving targets and long-time monitoring of the variational scene over the region of interest. As a new imaging mode, more and more scholars and agencies have focused on the video Synthetic Aperture Radar (SAR) due to it can provide continuous surveillance over the region of interest. The spaceborne video SAR has the corresponding advantages over the spaceborne SAR image system and the optical video system. The working principles, imaging algorithm, and application method of spaceborne video SAR have been proposed in this chapter. First of all, a theoretical System of spaceborne video SAR has been constructed. The operation and application mode have also been defined. Some key performances have been discussed. To meet the demand for video SAR applications, one imaging algorithm has been proposed for dealing with the spaceborne video SAR data. Experiments on simulated data show that the algorithm was effective.


**Keywords:** video SAR, working principles, imaging algorithm, moving target detection, parameter estimation

## 1. Introduction

Video SAR (synthetic aperture radar) is a new imaging mode that can provide continuous surveillance over a region of interest [1]. Compared with traditional SAR imaging, the SAR image stream of spaceborne video SAR is acquired by rapid imaging in a short time. Due to its dynamic information acquisition ability, video SAR is more suitable for the observation of moving targets and time-varying scenes [2–8]. The main work and contributions of this chapter are summarized as follows.

1. A theoretical system of spaceborne video SAR has been constructed. Based on the traditional spaceborne SAR system, the concept of spaceborne video SAR was introduced. The operation and application mode have also been defined. Some key performances, such as resolution, duration of the different operation times, and division method of the raw data have been discussed.

2. To meet the demand for video SAR applications, such as high accuracy and efficient computation. One imaging algorithm has been proposed for dealing with the spaceborne video SAR data. The image formation algorithms can avoid the duplication of processing, to improve the computation efficiency. Finally, experiments on simulated data show that the proposed algorithms were effective.

This chapter has broken through some key technologies in the construction and application of spaceborne video SAR system. The research result can be used as advice to build a spaceborne video SAR system.

## 2. The theory of spaceborne video SAR

### 2.1 SAR imaging

Synthetic aperture radar is a two-dimensional high-resolution imaging radar, whose high resolution in the range direction is achieved by transmitting LFM signal followed by matched filtering. The azimuth high resolution is achieved by using the relative motion between the radar and the target to form an equivalent large aperture [9, 10].

### 2.2 Spaceborne video SAR

Broadly speaking, video is generally defined as a number of linked images played continuously at a certain frequency, which forms a moving image.

Narrowly defined video is generally used in movies or television, and refers to continuous image changes of more than 24 frames per second, according to the principle of visual transient, the human eye cannot distinguish a single static picture, looks like a smooth continuous visual effect, such a continuous picture is also called video.

The U.S. Defense Advanced Research Projects Agency (DARPA) has made a preliminary definition of video SAR: the technology that can reflect a series of SAR images of continuous changes of a target or scene displayed at a fixed frame rate is called video SAR [6].

#### *2.2.1 The operating mode of spaceborne video SAR*

In the application of spaceborne video SAR, multiple SAR imaging of the same scene is mainly realized, while the effective observation time is increased as much as possible. For airborne platforms, circular-track SAR is generally used to realize long-time observation of the scene, while for spaceborne platforms, the observation time can be effectively extended through reasonable orbit design to realize video observation.

1. GEO Video SAR

GEO SAR satellites can form a near-circular satellite trajectory to the earth through orbit design, providing the possibility of GEO circular-track video SAR, which can effectively extend the observation time while realizing the gaze on fixed scenes.

**Figure 1.**
*GEOSAR satellite circular trajectory video SAR subsatellite point.*

For geosynchronous orbit satellites, circular-track SAR for video imaging can be achieved by making the sub-satellite point trajectory circular. The design method is to control the north–south drift of the satellite to be equal to the east–west drift, the north–south drift is determined by the orbit inclination, while the east–west drift is determined by the eccentricity, and the ascending node determines the longitude of the circle track center. Thus, the circle-trace video observation can be realized by a reasonable orbit design. The sub-satellite point trajectory in the process of GEO circular-track video observation was shown in **Figure 1**.

2. Spotlight video SAR

To prolong the video SAR imaging time, the low-orbit satellite video SAR can work in the large-angle staring spotlight mode, in which the satellite uses the azimuth direction large-angle sweep capability to achieve a long time observation of the observation scene, and the conventional spotlight SAR improves the azimuthal resolution by extending the observation time, while the video SAR can achieve multiple imaging by reducing the azimuth resolution of a single image frame. Video imaging can be achieved by reasonably segmenting the echo data throughout the imaging time, and the geometric schematic of its imaging mode was shown in **Figure 2**.

3. Sliding spotlight video SAR

In order to increase the azimuthal width of the video imaging scene, the video SAR can also operate in sliding spotlight SAR, and the geometric schematic of its imaging mode was shown in **Figure 3**, in which the satellite uses its agile maneuvering capability to perform sliding beam imaging of the observed scene, and then performs attitude maneuvering after the first image is completed to complete the second frame of video imaging, and repeats the imaging until it is beyond the satellite's attitude maneuvering capability or observable range, thus forming a video SAR image.

### *2.2.2 The applications of spaceborne video SAR*

The spaceborne video SAR is actually a sequence of SAR images of the same target area with high update frequency, and the applications based on video SAR images mainly include the following aspects [11–13].

1. Multi-aspect observation of target

For the spaceborne video SAR, different video frames have different observation aspects of the target, so different video frames can achieve multi-aspect observation of

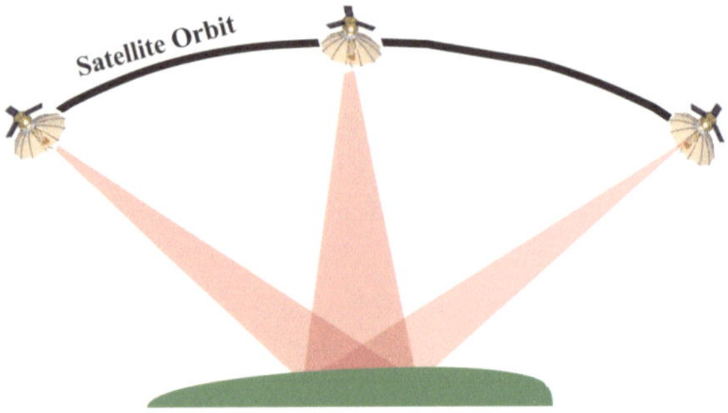


**Figure 2.**
*The geometric schematic of spotlight video SAR.*

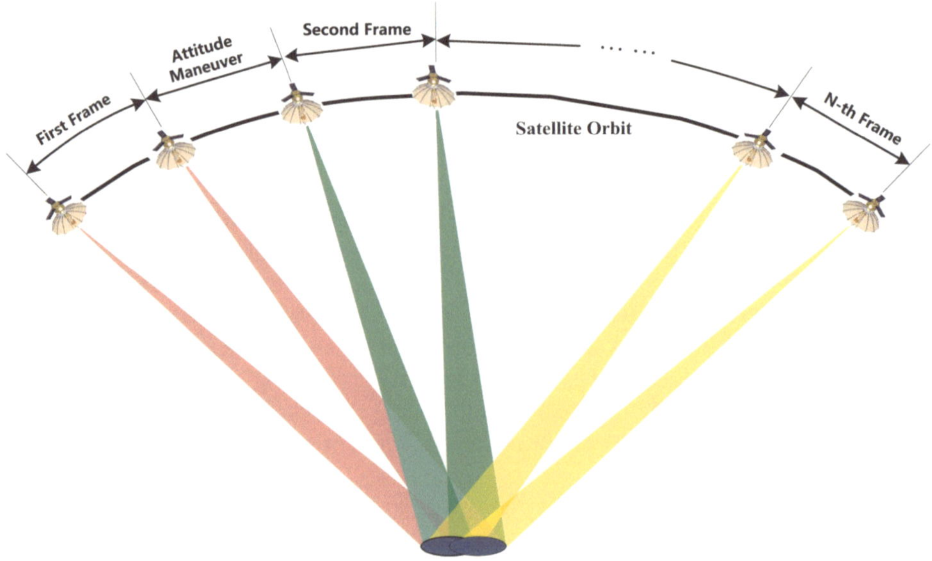


**Figure 3.**
*The geometric schematic of sliding spotlight video SAR.*

the target, and the SAR images of the target under different observation aspects can fully describe the characteristic information of the target, which is of great significance to the target identification and confirmation.

2. Suppressing the coherent speckle of SAR images.

Coherence speckle refers to the subechoes of multiple scattering points superimposed or eliminated with the same phase in certain resolution units, which makes dotted bright or dark areas appear in SAR images. The traditional coherent speckle suppression method uses multi-look processing to achieve the non-coherent superposition of multi-look images. The effective suppression of coherent speckle can also be achieved in video SAR products by non-coherent superposition of multi-frame images.

3. Continuous monitoring of scenarios and targets [14–18]

The existing SAR moving target detection technology does not achieve continuous video monitoring of hotspot areas, and the ATI or DPCA-based moving target detection has the problems of minimum detection speed and blind speed, and the estimation of target motion parameters also has the problem of ambiguity, which brings greater challenges to the localization and imaging of moving targets. Video SAR effectively extends the information in the time dimension, and the detection of moving targets and estimation of motion parameters can be achieved by using the change information between frames. The video SAR products after locating and imaging moving targets can intuitively display the motion information such as position, velocity, and motion trend of moving targets in stationary scenes.

### *2.2.3 Analysis of imaging duration of spaceborne video SAR*

The imaging duration of the spaceborne video SAR is mainly constrained by the following factors, and the minimum value of the video imaging duration determined by these factors is the imaging duration of the spaceborne video SAR.

1. Incidence angle constraint

The variation of the incidence angle affects the ground range resolution, and the incidence angle also affects the radar observation distance. In general, the backscattering cross section decreases with the increase of the incidence angle. From the perspective of energy return, the incidence angle should be chosen as small as possible within the applicable range of scattering theory, but from the perspective of application requirements, different application requirements require different incidence angles. The difficulty of system implementation will increase after the incidence angle is extended. Considering these factors, the constraint of incidence angle needs to be satisfied in the process of spaceborne video SAR imaging.

2. The beam azimuth sweeping capability constraint

The imaging duration of spaceborne video SAR is also constrained by the beam sweeping capability of the satellite. For phased-array antennas, the antenna gain decreases seriously during the large-angle sweeping process, and the dispersion effect will occur at the same time, resulting in the degradation of imaging quality, while the reflector antenna has the characteristics of stable antenna pattern and high gain during the large-angle sweeping process. For the reflector antenna, the azimuth sweeping capability is mainly limited by the maneuverability of the agile platform. The sweeping capability determines the duration of the video imaging, and the larger the sweeping angle, the longer the video imaging time within the incidence angle meets the observation requirements.

3. Squint angle constraint

The coupling between range and azimuth direction is serious in the process of SAR imaging processing when the squint angle is large, which brings difficulties to the imaging processing, and the existing imaging processing algorithm has the maximum squint angle limitation if a good focusing effect is to be achieved. Therefore, the requirement of squint angle during imaging processing is also one of the constraints on the duration of video SAR imaging.

## 3. Image formation algorithm of spaceborne video SAR

The main two modes of video SAR implementation by LEO satellites are spotlight and sliding spotlight. Since there is no overlap of data between adjacent frames in sliding spotlight mode, the imaging algorithm in this mode is the same as the traditional imaging algorithm in sliding spotlight mode. In contrast, for the spotlight mode, there is overlap between adjacent video frames, and to avoid repeated operations of overlapping data, the imaging algorithm applicable to spaceborne video SAR needs to be studied. In this section, a video SAR imaging algorithm is proposed for the key technical problems to be solved in the process of video SAR imaging, and the simulation is verified.

### 3.1 Echo data segmentation method

Typically, the spaceborne radar is operated in the spotlight mode for an extended period of time while taking video. The conventional spotlight SAR imaging mode achieves the high azimuth resolution through lengthening the synthetic time, while in the video mode the spotlight raw data was divided into pieces to form the video frames. The image geometrical mode of spaceborne video SAR based on equivalent squint range mode was shown in **Figure 4**, in which $L_s$ is the distance the radar moves throughout the video imaging progress. $\beta$ is the largest synthetic angle, $l_s$ is the distance the radar moves of a video frame. $\theta_i$ is the synthetic angle of a video frame, while $\theta_{ci}$ is the squint angle of the video frame. The equivalent velocity of the radar is $v_r$.

The Doppler bandwidth of the ith video frame can be expressed as follows.

$$Ba_i = fd_{\mathrm{b}} - fd_{\mathrm{a}} \tag{1}$$

Where $fd_{\mathrm{a}}$ and $fd_{\mathrm{b}}$ are the Doppler frequency at the start and end time of the ith video frame.

$$fd_{\mathrm{a}} = -\frac{2v_{\mathrm{r}}\cos\left(\theta_{ci} - \frac{\theta_i}{2}\right)}{\lambda} \tag{2}$$

$$fd_{\mathrm{b}} = -\frac{2v_{\mathrm{r}}\cos\left(\theta_{ci} + \frac{\theta_i}{2}\right)}{\lambda} \tag{3}$$

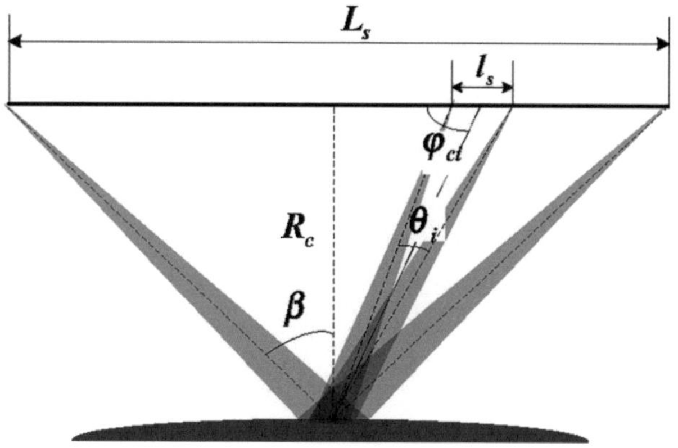


**Figure 4.**
*The image geometrical mode of spaceborne video SAR.*

The Doppler bandwidth of the ith video frame can be presented as follows.

$$Ba_i = fd_b - fd_a \approx \frac{2v_r\theta_i \sin\varphi_{ci}}{\lambda} \tag{4}$$

The azimuth resolution can be expressed as follows.

$$\rho_{ai} = \frac{v_g}{Ba_i} = \frac{\lambda v_g}{2v_r\theta_i \sin\varphi_{ci}} \tag{5}$$

Then the distance the radar moves of the ith video frame can be derived as follows.

$$L_i = \frac{\lambda v_g R_0}{2v_r\rho_{ai}\sin^3\varphi_{ci}} = \frac{\lambda v_g\left(R_0^2 + v_r^2 t_i^2\right)^{3/2}}{2v_r\rho_{ai}R_0^2} \tag{6}$$

Where $t_i$ is the middle time of the ith video frame.

In video SAR the synthetic aperture time to achieve the desired azimuth resolution typically exceed the frame period. As a result, there can be a significant overlap in the collected phase history used to form consecutive images in the video. **Figure 5**

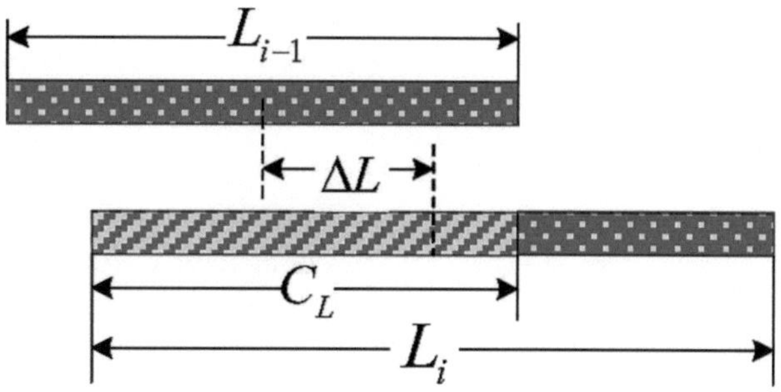


**Figure 5.**
*The overlap between adjacent frames of video SAR.*

illustrates the overlap between adjacent frames of video SAR, where $L_i$ is the synthetic aperture length of the ith video frame, $C_L$ is the overlap length of the adjacent frames. $\Delta L$ is the distance between adjacent frames.

The overlap length of the adjacent frames can be presented as follows.

$$C_{\mathrm{L}} = \frac{L_i}{2} + \left(\frac{L_{i-1}}{2} - \Delta L\right) = \frac{(L_i + L_{i-1})}{2} - \Delta L \tag{7}$$

Then the overlap rate can be expressed as follows.

$$\alpha_i = \frac{1}{2} + \frac{\lambda v_{\mathrm{g}}\left(R_0^2 + v_{\mathrm{r}}^2(t_i - T_{\mathrm{f}})^2\right)^{3/2} - 2T_{\mathrm{f}} v_{\mathrm{r}}^2 R_0^2 \rho_{ai}}{2\lambda v_{\mathrm{g}}\left(R_0^2 + v_{\mathrm{r}}^2 t_i^2\right)^{3/2}} \tag{8}$$

### 3.2 Spaceborne video SAR imaging based on BP algorithm

In the spaceborne video SAR system, the low-orbit satellite works in the spotlight mode, the video imaging is realized through the reasonable segmentation of the echo data, and the key technical problems to be solved in the spaceborne video SAR imaging mainly include as follows [19–26].

1. Large squint angle problem of some data frames

Since the LEO satellite works in the spotlight mode, some data frames have large squint angles, and there are serious range-azimuth coupling and range cell migration, which will lead to serious degradation of image quality by the traditional approximation method.

2. Data overlap of adjacent video frames

From the above analysis, it can be seen that in the spaceborne video SAR, since the synthetic aperture time is larger than the frame period, there is a large overlap of data between adjacent frames, and processing each frame individually will lead to repeated operations of overlapping data between adjacent frames, which will greatly reduce the operation efficiency.

3. Real-time problem

The spaceborne video SAR imaging needs to provide real-time or quasi-real-time video frame images, which requires the algorithm to have high computing efficiency, and the possibility of parallel computing needs to be explored based on the application of advanced computing hardware equipment.

A fast BP (Back-Projection) algorithm, which can be implemented in parallel, is proposed to address the imaging characteristics of space-based video SAR and the key technical problems to be solved. As an accurate time-domain algorithm, the BP algorithm avoids the geometric approximation, so it can well solve the problems of imaging under a complex distance model and the serious coupling between range and azimuth direction in the case of large squint angle in spaceborne video SAR [27].

In the application of the BP algorithm, the azimuth resolution increases with the increase of coherent cumulative pulse number, so the spaceborne video SAR imaging based on the BP algorithm can effectively avoid the repetitive operation of overlapping data of adjacent frames. The current frame can be imaged with the operation result of overlapping data of the previous frame, thus effectively improving the operation efficiency. At the same time, the sub-aperture division can further improve the operation speed of the BP algorithm, and the sub-aperture based processing can also realize parallel computing, which can effectively ensure the real-time or quasi-real-time output of video SAR images.

The process of spaceborne video SAR imaging based on the BP algorithm is shown in **Figure 6**, firstly, the echo signal is divided into several sub-apertures, and in order to achieve avoiding repeated operations through sub-aperture superposition, the frame period is required to be an integer multiple of the sub-aperture length, if the synthetic aperture length of a single video frame is $L$, and it is divided into $N$ sub-apertures, the length of each sub-aperture is $L_{\text{sub}} = L/N$. The imaging of each sub-aperture can be calculated in parallel to the image of each sub-aperture can be computed in parallel to generate a low-resolution image, and finally, the full-resolution image of a single frame can be obtained by sub-aperture synthesis.

The range pulse compression signal in a single sub-aperture is as follows.

$$s_{\text{BM}(ij)}(\tau, \eta) = s_{\text{BO}(ij)}(\tau, \eta) \otimes s^*_{\text{BO}(ij)}\left[\left(\text{T}_{\text{c}(ij)} - \tau\right), \eta\right] \tag{9}$$

Where $\tau$ is the range time, $\eta$ is the azimuth time, $\text{T}_{\text{c}(ij)}$ is the time delay of the reference point, since the echo time delay is difficult to coincide with the sampling point, the distance to the pulse compressed signal should be interpolated, and the interpolated signal is as follows.

$$S_{\text{BUM}(ij)}(t, \eta) = \sum_{|t - n\Delta\tau| \le N_s} s_{\text{BM}(ij)}(n\Delta\tau, \eta) h_{\text{w}}(t - n\Delta\tau) \tag{10}$$

Where $\tau = n\Delta\tau$ is the sampling point before interpolation, $2N_s$ is the length of interpolation kernel, $h_{\text{w}}(t)$ is the interpolation kernel function after window sharpening. Since the processed echo data has been demodulated, the echo phase compensation is needed to achieve coherent accumulation, and the result after the echo phase compensation is as follows.

$$S_{\text{BCUM}(ij)}(t, \eta) = S_{\text{BUM}(ij)}(t, \eta) \exp\left(j2\pi f_c t\right) \tag{11}$$

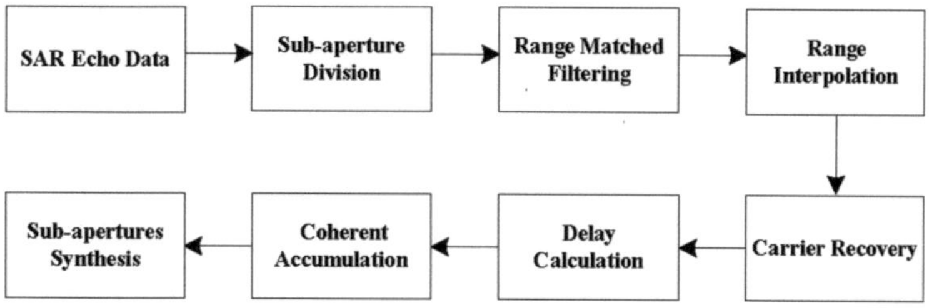


**Figure 6.**
*Flow diagram of one frame SAR image formation.*

The time delay between the target point and each radar position within the sub-aperture is as follows.

$$t_{ij}(n\Delta\eta) = R_{\mathrm{bi}(ij)}(n\Delta\eta)/c \tag{12}$$

Where $\eta = n\Delta\eta$ is the azimuth sampling point, then the imaging result of the target $\mathrm{P}_{ij}$ in the *kth* sub-aperture is as follows.

$$f^{k}_{\mathrm{B}(ij)}(\alpha_i, \delta_j) = \sum_{n=R}^{S} S_{\mathrm{BCUM}(ij)}\left[t_{ij}(n\Delta\eta), n\Delta\eta\right] \tag{13}$$

Where $R\Delta\eta$ is the starting time of the *kth* sub-aperture echo data, and $S\Delta\eta$ is the end time of the *kth* sub-aperture echo data, At this point, the low-resolution imaging result for the *kth* sub-aperture $f^{k}_{\mathrm{B}}(\alpha, \delta)$ is obtained by traversing each grid in the scene.

In the imaging process, the sub-apertures are calculated in parallel, which can effectively improve the computing efficiency. Finally, the low-resolution images of sub-apertures are coherently added to obtain the full-resolution image of the *ith* video frame as follows.

$$F_i(\alpha, \delta) = \sum_{k=1}^{N} f^{k}_{\mathrm{B}}(\alpha, \delta) \tag{14}$$

### 3.3 Simulation

To validate the video SAR algorithm, a point target simulation is carried out based on the parameters in **Table 1**.

The scene size is $2 \times 2\,\mathrm{km}$. There are 25 point targets in the simulated scene, the initial moment is arranged uniformly by $5 \times 5$, the first and fifth columns and the first and fifth rows of the point targets are stationary, and the coordinates of the target in the center of the scene are assumed to be $(0, 0)$. The motion parameters of the remaining targets are shown in **Table 2**.

| Parameters | Values |
|---|---|
| Earth Radius (km) | 6378 |
| Orbital Eccentricity | 0 |
| Orbital Inclination (°) | 20 |
| Orbital Height (km) | 567 |
| Longitude of Ascending Node (°) | 300 |
| Radar Center Frequency (GHz) | 9.6 |
| Transmitted Pulse Duration (us) | 30 |
| Transmitted Signal Bandwidth (MHz) | 150 |
| Azimuth Resolution (m) | 1 |
| Video Frame Rate (Hz) | 5 |

**Table 1.**
*Parameters for simulation.*

| 目标 | 坐标 | $v_r/(\mathbf{km{\cdot}h^{-1}})$ | $a_r/(\mathbf{km{\cdot}h^{-2}})$ | $v_a/(\mathbf{km{\cdot}h^{-1}})$ | $a_a/(\mathbf{km{\cdot}h^{-2}})$ |
|---|---|---|---|---|---|
| $S(2,2)$ | $(-0.5, 0.5)$ | −15 | 0 | 0 | 0 |
| $S(2,3)$ | $(0, 0.5)$ | −10 | 0 | 0 | 0 |
| $S(2,4)$ | $(0.5, 0.5)$ | 0 | 0 | 0 | 50 |
| $S(3,2)$ | $(-0.5, 0)$ | 0 | 0 | 120 | 0 |
| $S(3,3)$ | $(0, 0)$ | 0 | 0 | 200 | 0 |
| $S(3,4)$ | $(0.5, 0)$ | 0 | −5 | 0 | 0 |
| $S(4,2)$ | $(-0.5, -0.5)$ | 15 | 0 | 120 | 0 |
| $S(4,3)$ | $(0, -0.5)$ | 10 | 0 | 200 | 0 |
| $S(4,4)$ | $(0.5, -0.5)$ | 0 | −5 | 0 | 50 |

**Table 2.**
*Motion parameters of moving targets.*

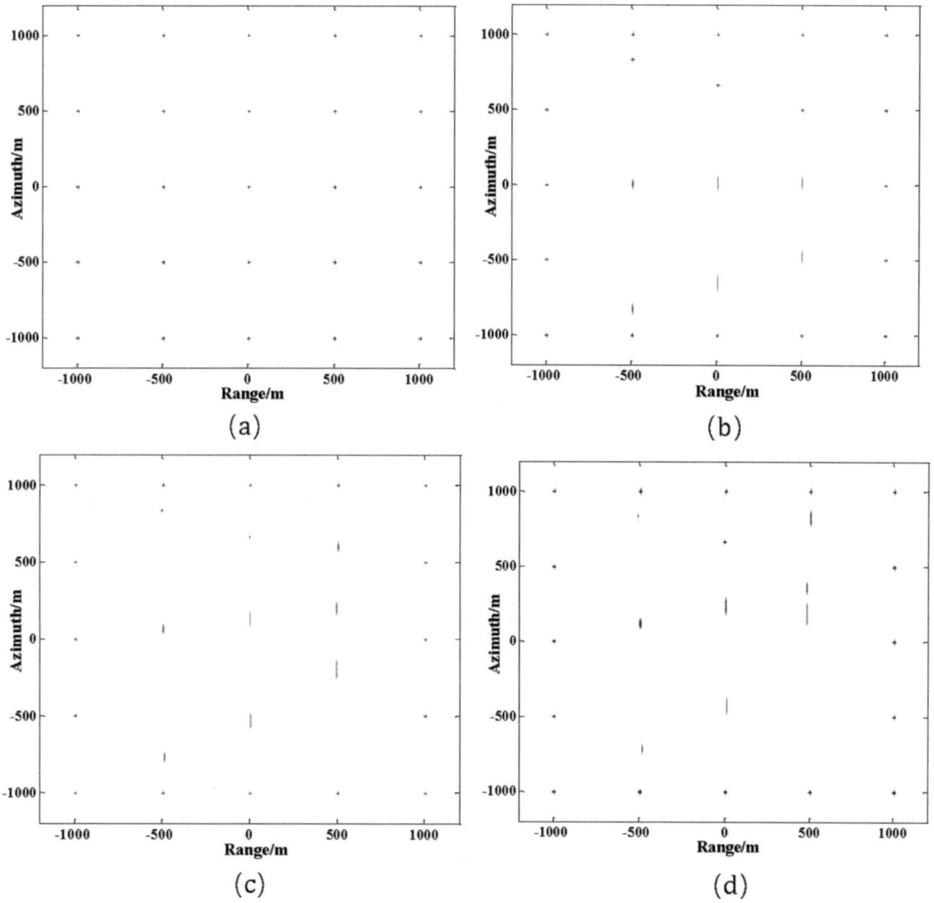


**Figure 7.**
*Image formation result of spaceborne video SAR. (a) Stationary target imaging results (b) The first frame of the video SAR (c) The 10th frame of the video SAR (d) The 18th frame of the video SAR.*

The results of spaceborne video SAR simulation are shown in **Figure** 7 with a frame rate of 5 Hz, where **Figure** 7**(a)** shows the imaging results of 25 points when the target is stationary in the initial state, and it can be seen that the target achieves a good focusing effect at each point when the target is stationary. **Figure** 7**(b)** shows the imaging results of the 1st frame of the spaceborne video SAR, and the comparison of the imaging results of $S(2,2)$, $S(2,3)$ and $S(2,4)$ show that the larger the velocity in the range direction, the larger the offset of the azimuth direction of the target, and the velocity in the range direction and the acceleration in the azimuth direction have less influence on the azimuth spreading of the target. The comparison of the imaging results of $S(3,2)$, $S(3,3)$ and $S(3,4)$ show that the larger the azimuth velocity of the target, the more serious the azimuth spread of the target, and the range acceleration is also the main cause of the azimuth spread of the target. From the imaging results of $S(4,2)$, $S(4,3)$ and $S(4,4)$, it can be seen that when the target has both azimuth velocity, range velocity and range acceleration, the azimuth image of the target is both shifted and defocused. **Figure** 7**(c)** and **(d)** show the imaging results of frame 10 and frame 18 of the spaceborne video SAR respectively. From the imaging results of $S(2,4)$, $S(3,4)$ and $S(4,4)$, it can be seen that the azimuth velocity and range velocity of the three targets gradually become larger and the azimuth spreading and shifting become larger as the time increases. It can be seen that the imaging results of spaceborne video SAR correctly reflect the motion information of the targets, which can provide the basis for the subsequent motion target detection, motion parameter estimation and repositioning and imaging of the moving targets based on SAR video.

## 4. Conclusions

In this chapter, a general definition of spaceborne video SAR is given first, and three operating modes and possible application directions of spaceborne video SAR are proposed for the demand of long-time observation. The imaging duration of spaceborne video SAR is mainly affected by the incidence angle, azimuth sweeping capability and the maximum squint angle allowed by the imaging process. The analysis shows that for low-orbit satellites, the angle of incidence is the main factor limiting the duration of video SAR. Then a parallel computable video SAR imaging algorithm based on sub-aperture division is proposed for the three types of key technical problems to be solved in spaceborne video SAR imaging, and computer simulation is conducted to verify the results. The simulation results show that the video imaging results correctly reflect the motion of the target and can provide the basis for the motion target detection, parameter estimation, and repositioning and imaging based on SAR video. The research results of this chapter can provide suggestions and references for the construction and application of future spaceborne video SAR systems.

## Conflict of interest

The authors declare no conflict of interest.

## Author details

Jian Liang* and Liang An
Institute of Remote Sensing Satellite, China Academy of Space Technology, Beijing, China

*Address all correspondence to: liangjiancast@163.com

## References

[1] DARPA. Video synthetic aperture radar (ViSAR). Broad Agency Announcement. 2012:DARPA-BAA-12-41:1-51

[2] Well L, Doerry A, et al. Developments in SAR and IFSAR System and Technologies at Sandia National Laboratories. Vol. 2. Big Sky, MT, USA: IEEE; 2005. pp. 1085-1095

[3] Available from: http://www.sandia.gov/radar

[4] Damini A, Balaji B, Parry C. A video SAR mode for the X-band wideband experimental airborne radar. Proceedings of SPIE. 2010;**7699**(0E):1-11

[5] Moses RL, Joshua N. Recursive SAR imaging. In: Proceedings of the SPIE Algorithms for Synthetic Aperture Radar Imagery XV. Orlando, FL, USA: Ohio State University Department of Electrical and Computer Engineering Columbus OH USA. Vol. 6970. 2008. pp. 1-12

[6] Defense Advanced Research Projects Agency. Broad Agency Announcement: Video Synthetic Aperture Radar System Design And Development. DARPA, USA: Arlington; 2012

[7] Wallace HB. Development of a video SAR for FMV through clouds. Proceedings of SPIE. 2015;**9479**(0L):1-2

[8] Palma S, Wahlen A, Stanko S, et al. Real-time onboard processing and ground based monitoring of FMCW-SAR videos. In: The 2014 EUSAR. Vol. 3607. Berlin: IEEE; 2014. pp. 2-5

[9] Cumming IG, Wong FH. Digital Processing of Synthetic Aperture Radar Data Algorithms and Implementation. London: Artech House Inc; 2005

[10] Curlander JC, McDonough RN. Synthetic Aperture Radar: Systems and Signal Processing. NY, USA: John Wiley and Sons, Inc.; 1991

[11] Elachi C. Spaceborne Radar Remote Sensing: Applications and Techniques. New York: The Institute of Electrical and Electronic Engineers, Inc; 1987:33-42

[12] Damini A, Mantle V, Davidson G. A new approach to coherent change detection in video SAR imagery using stack averaged coherence. In: The 2013 IEEE Radar Conference. Vol. 5794. Ottawa, ON, Canada: IEEE; 2013. pp. 13-17

[13] Carrara WG, Goodman RS, Majewski RM. Spotlight Synthetic Aperture Radar: Signal and Processing Algorithms. Norwood, MA, USA: Artech House; 1995

[14] Kirscht M. Detection and velocity estimation of moving objects in a sequence of single-look SAR images. Physical Review A. 1996;**1**(5):333-335

[15] Ouchi K. On the multi look images of moving targets by synthetic aperture radars. IEEE Transactions on Antennas and Propagation. 1985;**8**(33):823-827. DOI: 10.1109/TAP.1985.1143684

[16] Kirscht. Method of detecting moving objects and estimating their velocity and position in SAR images United States Patent, 2005.10.04:US6952178B2

[17] Kirscht M. Detection and imaging of arbitrarily moving targets with single-channel SAR. Radar, Sonar and Navigation, IEEE Proceedings. 2003; **150**(1):7-11. DOI: 10.1109/RADAR.2002.1174697

[18] Yang J, Liu C, Wang Y. Imaging and parameter estimation of fast-moving targets with single-antenna SAR. IEEE Geoscience and Remote Sensing Letters. 2014;**11**(2):529-533. DOI: 10.1109/LGRS.2013.2271691

[19] Zhao S, Chen J, Yang W, et al. Image formation method for spaceborne video SAR. In: IEEE 5th Asia-Pacific Conference on Synthetic Aperture Rad. Singapore: IEEE; 2015. pp. 148-151

[20] Linnehan R, Miller J, Bishop E. An autofocus technique for video-SAR. Proceedings of SPIE. 2013;**8746**(08):1-10

[21] Miller J, Bishop E, Doerry A. Applying stereo SAR to remove height-dependent layover effects from video SAR imagery. Proceedings of SPIE. 2014; **9093**(3A):1-10

[22] Hawley RW, Garber WL. Aperture weighting technique for video synthetic aperture radar. Proceedings of SPIE. 2011;**8051**(07):1-7

[23] Jinping S, Zhifeng Y, Wenb H. A new subaperture chirp scaling algorithm for spaceborne spotlight SAR data focusing. In: The 2007 IEEE Radar Conference. Vol. 5794. London: IEEE; 2013. pp. 13-17

[24] Kaizhi Wang, Xingzhao Liu. Squint-spotlight SAR imaging by sub-band combination and rage-walk removal. Geoscience and Remote Sensing Symposium. Vol. 8742:(2) Alaska: IEEE; 2004. 3930-3933.

[25] Jin L, Liu X, Wang J. Adaptive subaperture approach for spotlight SAR azimuth processing. In: Geoscience and Remote Sensing Symposium. Boston: IEEE. Vol. 2808:(3) 2008: 1292-1295.

[26] Wang W, Ma X. A novel data preprocessing method for resolving doppler ambiguity of spaceborne spotlight SAR. In: Geoscience and Remote Sensing Symposium. Munich: IEEE; Vol. 978:(1) 2012:5-8.

[27] Miller J, Bishop E, Doerry A. An application of backprojection for Video SAR image formation exploiting a subaperture circular shift register. Proceedings of SPIE. 2013;**8746**(09):1-14

Chapter 4

# Evolution of Attacks on Intelligent Surveillance Systems and Effective Detection Techniques

*Deeraj Nagothu, Nihal Poredi and Yu Chen*

## Abstract

Intelligent surveillance systems play an essential role in modern smart cities to enable situational awareness. As part of the critical infrastructure, surveillance systems are often targeted by attackers aiming to compromise the security and safety of smart cities. Manipulating the audio or video channels could create a false perception of captured events and bypass detection. This chapter presents an overview of the attack vectors designed to compromise intelligent surveillance systems and discusses existing detection techniques. With advanced machine learning (ML) models and computing resources, both attack vectors and detection techniques have evolved to use ML-based techniques more effectively, resulting in non-equilibrium dynamics. The current detection techniques vary from training a neural network to detect forgery artifacts to use the intrinsic and extrinsic environmental fingerprints for any manipulations. Therefore, studying the effectiveness of different detection techniques and their reliability against the defined attack vectors is a priority to secure the system and create a plan of action against potential threats.

**Keywords:** intelligent surveillance systems, internet of video things (IoVT), multimedia forgery, environmental fingerprints, forgery detection, DeepFake detection

## 1. Introduction

The modern smart city infrastructure has advanced by integrating multimedia-based information input and the development of an edge computing paradigm [1, 2]. An increase in visual and auditory input from the deployed sensors has enabled multiple network layer-based processing of incoming information. While most of the intelligent infrastructure depends on a cloud computing-based architecture [3], edge computing has been attracting more and more attention to meet the increasing challenges in terms of scalability, availability, and the requirements of instant, on-site decision making [4–6]. Advancements in artificial intelligence (AI) have equipped the edge computers to process the incoming multimedia feed and deploy recognition and detection software. Machine learning (ML)-based models such as object detection, tracking, speech recognition, and people identification are commonly deployed to enhance the security in infrastructure and private properties [7]. With an increase in such technological advancements, the system's reliability has also exponentially increased where the trust factor established on the system is directly depending on the information retrieved by the multimedia sensor nodes [8]. The edge devices are enhanced with multi-node communication and equipped with Internet connections to provide continuous functionality and security services.

Due to their significance in infrastructure security and functionality, edge computing devices are commonly targeted through networked attacks through Wi-Fi and RF links [9]. The devices are compromised through malicious firmware updates [10] and result in creating a backdoor with admin privileges. The perpetrators then control the device Input/Output (IO) and compromise the network and home security. Specifically, visual layer attacks are developed to manipulate the visual sensor in edge devices and create a false perception of live events monitored by the control station. Simple frame manipulation such as frame duplication or shuffling allows the perpetrator to mask the original frame, where the security of the infrastructure can be easily compromised [11]. There is also no evidence of crimes without the surveillance recordings, and it falters the need for such security devices. Along with the visual layer, the audio channel of the edge nodes is equally targeted. Modern home security is enabled with voice commands and a home assistant system that functions based on the voice commands received. The audio devices are equipped with voice-based home assistant computers and Voice Over IP (VoIP) surveillance recorders. The attackers can target the audio channel through hidden voice commands, control the system, or completely mask the audio channel with noise to disable its functionality [12].

As the ML-based models have enhanced the surveillance system's capabilities, it has also resulted in the development of frame manipulation attacks. Beginning with the traditional copy-move style forgery attacks in spatial regions of a frame, modern deep learning (DL) has enabled generative networks capable of creating a frame based on the user's input. Adversarial networks have rendered some ML models useless due to their targeted attack to disable their functionality. General adversarial networks (GANs) have created DeepFakes, which have become one of the most challenging problems in current multimedia forgery attacks [13]. DeepFake is trained to function in low computing systems such as edge devices and result in manipulations such as Face Swaps, Facial Re-enactments, and complete manipulation of the targeted person's movements resulting in a very realistic media output [14]. It is clear that both the visual and auditory channels require robust security measures and reliable authentication schemes to detect such malicious attacks and secure the network [15, 16].

Advancements in forgery attacks have always been countered with detection schemes. Traditional frame forgery attacks were first detected using watermark technology and compression artifacts [17]. However, when the edge device is compromised, the frames are manipulated at the source level, creating watermarks on false frames. Similarly, with DeepFake being developed, its counterpart detection schemes were also trained. The first stages of DeepFakes carry visual artifacts like face recordings without any eye blinking or face warping artifacts [18]. Still, with more training data and better networks, DeepFakes have evolved to a point where it is almost not distinguishable from real images [19]. Although the technology itself has its own merits when ethically used in the field of medical and entertainment, perpetrators can always use the DeepFake technology with malicious intent without a reliable detector. It is an ongoing effort to create a reliable detection scheme to clearly distinguish between real and fake.

This chapter provides an overview of the evolution of multimedia-based attacks to compromise the edge computing nodes such as surveillance systems and their counterpart forgery detection schemes. The essential features required by a reliable detection system are analyzed and a framework using an environmental fingerprint is introduced that has proven to be effective against such attacks.

## 2. An overview of audio-visual layer attacks

The networked edge devices are commonly deployed through Wi-Fi or RF links in a private network. The primary means of hijacking a secure device is through network layer attacks where the communication between the devices is intercepted and modified [20]. This allows the source and the destination to believe that the information exchange was secure, while a perpetrator alters the intercepted message as required. Malicious firmware is updated through direct physical access to the USB interface or remote web interface, which allows a perpetrator to gain admin privileges to the edge devices. Some devices are sold through legitimate channels with malicious firmware pre-installed [10]. With complete access to the visual and audio sensor nodes, the attacker can manipulate the media capturing module itself, making the network-level security measures compromised.

Surveillance systems are the most targeted edge devices due to their importance and access medium [11]. Network attacks like Denial of Service (DoS) can disable the network connections of the devices and negate their purposes. Common admin mistakes like using the default credentials on the networks and devices login are primary reasons for backdoor entry. Once the device or the network is compromised, the attacker typically encodes the trigger mechanism into the system. This allows the perpetrator to remotely trigger the selected attack based on remote commands without re-accessing the device. Malicious inputs can be encoded into the multimedia encoding scheme of the edge device. Trigger methods like QR-code-based input to the video recording interpret the command differently [21], face detection-based trigger [22], and hidden voice commands through the audio channels [12] are a few examples of how an attack can be remotely controlled. Wearable technologies like Google Glass are also affected through the backdoor firmware, where the QR-code-based input was used to hack the device [23].

With remote trigger mechanisms, a device can be controlled to manipulate the incoming media signal. Face detection software can be re-programmed to blur selected faces and car plate registrations or disable certain functionality like detecting prohibited items like guns [9]. Popular Xerox scanners and photocopiers were hacked to manipulate the contents of the documents that are scanned and insert random numbers instead of actual data [24]. Surveillance cameras with Pan-Tilt-Zoom (PTZ) capabilities can be controlled to re-position the cameras so that the number of blind spots is increased in a surveillance area [25]. Audio Event Detectors (AED) are commonly deployed in surveillance devices to raise the alarm based on suspicious audio activity or in-home assistant devices to detect the wake commands. Still, the AED system can be directly targeted using the hidden voice commands to interpret its input falsely [12, 26]. Using the adversarial networks, popular ML models on edge devices are targeted so that the input itself can be modified [27]. Frame-level pixel manipulations are made to confuse the ML models and result in the false categorization of object recognition models [28]. A wearable patch is trained to target the person identification ML model, which can be worn by a perpetrator in the form of a t-shirt and escape the identification module [29].

Access to the multimedia sensor nodes can result in many variants of visual and audio layer attacks. To study the effective detection methods, we first narrow the video frame manipulation and audio overlay attacks commonly designed to target the edge-based media input such as surveillance devices and online conferencing technologies.

### 2.1 Frame manipulation attacks

Video recordings used for temporal correlation of the live events are primarily targeted using frame shuffling or duplication attacks [30]. The perception of live events is affected, which disables the effectiveness of live monitoring [31]. Adaptive replay attacks are designed such that the frame duplication attack can adapt to the changes in the environments such as light intensity variations, object displacement, and camera alignments. With adjusting frame masking, the operator in the monitoring station cannot distinguish between the real and fake images since the duplicated frames are originally copied from the same source camera [22]. The effect of source device identification and watermarking technique is negated since the frames originated in the same camera. **Figure 1** represents a frame replay attack where the attack is triggered remotely by either a QR-code or face detection module, and the resulting frame is masked with a static background [21, 32].

Spatial manipulation of a frame includes changes to the pixels like object addition or deletion, while the static frame is maintained. Frame-level manipulations are commonly made to deceive the viewer with the presence of a subject [33, 34]. The figure shows the spatial manipulation of the video frame.

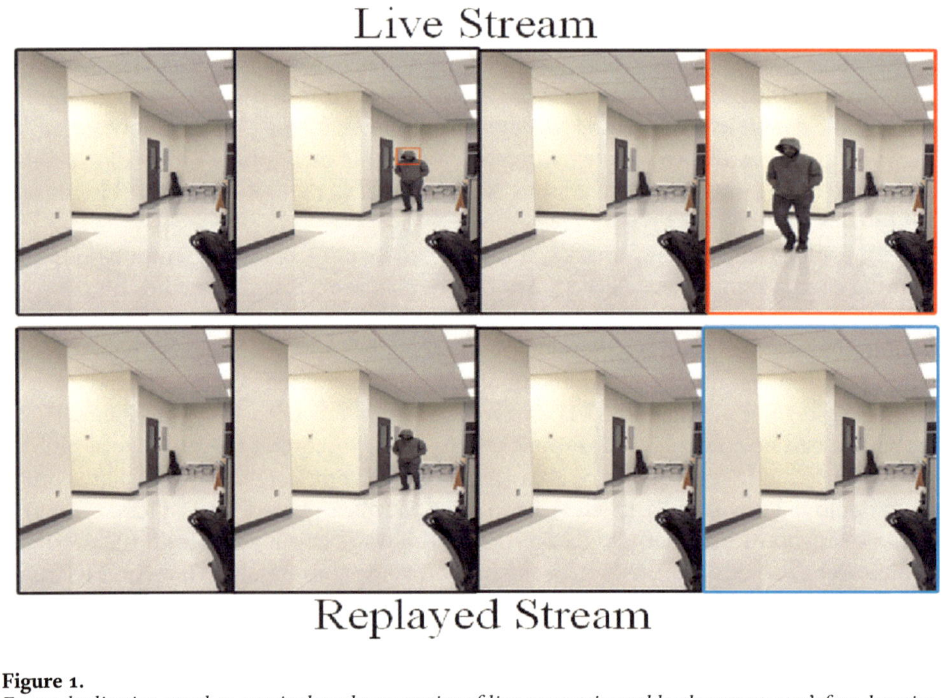


**Figure 1.**
*Frame duplication attack to manipulate the perception of live events triggered by the perpetrator's face detection.*

*DOI: http://dx.doi.org/10.5772/intechopen.105958*

### 2.2 Audio masking and overlay

Most edge nodes are equipped with audio recording capabilities making them a target for forgery attacks [3]. Every household is equipped with surveillance cameras, home assistants, and edge devices capable of two-way communications. The AED module is responsible for wake command detection or event detection based on audio like gunshot sounds. The input audio sensor nodes are disabled by compromising the AED module by replacing the actual event with the quiet static noise. The input is also affected by adding additional white noise to disrupt the AED module [26].

### 2.3 DeepFake attack

DeepFake attacks developed using GAN architecture [13] have resulted in a large quantity of fake media generation. With enough training data available and the computation resources, the quality of the generated media keeps improving to a point where a person cannot distinguish between the real and fake media [35]. Although DeepFake technology has its application merits, any technology can cause more harm than good in the wrong hands. The developing software technologies have made it easier and more convenient for the generation of DeepFake media using their mobile phone.

A simple face manipulation software where two people can swap their facial landmarks originated in the form of mobile applications. Soon, advanced technologies were made to make the swap more realistic [36]. Many organizations and institutions rely on online conferencing solutions for their daily communications. Face-swapping technologies allow perpetrators to mimic a source facial landmark and duplicate their online personality [37]. However, with the capability to extract facial landmarks and skeletal features from a source subject, a new form of DeepFake emerged to project source movement on a targeted subject (**Figure 2**).

The facial re-enactment software [38] allows the model to extract the face landmark movements from a source subject. These landmarks are projected on a targeted

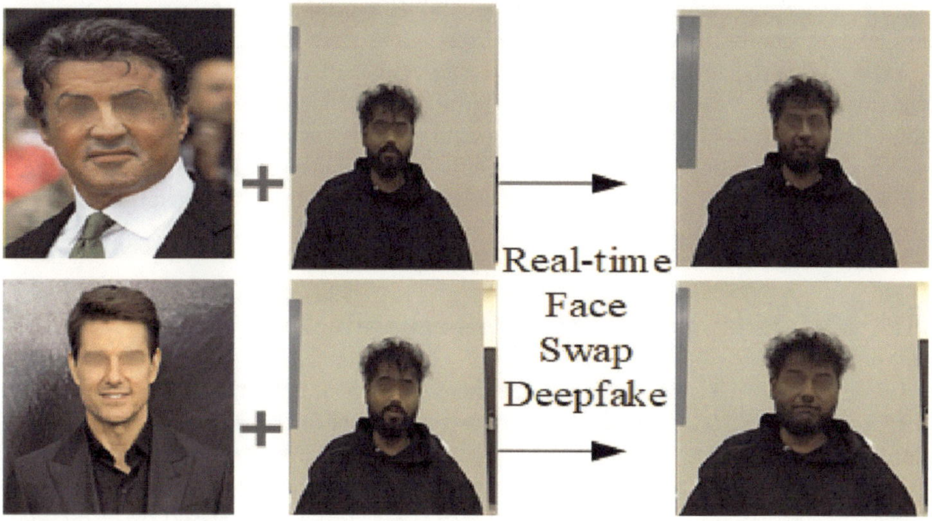


**Figure 2.**
*DeepFake Face Swap Attack to project a source face on a target.*

victim resulting in a media where the victim is projected to act out however the perpetrator wishes. Although the model was created to demonstrate the capabilities of deep learning models, it resulted in targeting politicians and celebrities to develop fake media. A GAN model is created where the source body actions are projected on a targeted person [39]. The model introduced resulted in creating an entertainment application, and it could also be alternatively used to frame a victim by forging their actions in surveillance media. The style-based transfer learning has enabled the GAN technology to create more realistic and indistinguishable output [19].

Introducing perturbations in real objects or images can cause edge layer object classifiers to make incorrect predictions, which could have serious repercussions. A study showed that making small changes in a stop sign could cause an object detector to wrongly classify it as a different object as depicted in 3(a) [40]. This phenomenon has been analyzed and the Fast Gradient Sign Method attack was proposed, which uses the gradient of the loss function of the classifier to construct the perturbations necessary to carry out the attack [41]. The attack begins by targeting an image and observing the confidence of the classifier in its predictions of the class. Next, the minimum perturbation that maximizes the loss function of the classifier is found iteratively. Using this method, the image can be manipulated such that incorrect classification is achieved without producing any discernible difference to the human eye as shown in 3(b). The Jacobian-based Saliency Map Attack [42] algorithm computes the Jacobian matrix of the CNN being used for object classification and produces a salient map. The map denotes the scale of influence each pixel of the image has on the prediction of the CNN-based classifier. The original image is manipulated in every iteration, such that the two most influential pixels, which are chosen from the saliency map, are changed. The salient map is updated in each iteration, and each pixel is changed only once. This stops when the adversarial image is successfully classified to the target label (**Figure 3**).

**Table 1** summarizes the multimedia attack techniques and their respective targeted systems. Along with video manipulation, audio is also equally targeted when creating realistic fake media. Paired with technology like facial re-enactment, DeepFake audio can create an illusion of a targeted person with manipulated actions. Software like Descript [43] can recreate source audio with training data for few as 10 minutes. Emerging technologies like DeepFake need a reliable detector that can distinguish between real and fake media to preserve security and privacy in the modern digital era. Due to the inconsistencies in earlier stages of DeepFake media, many detector modules were created to identify the artifacts introduced during media generation. However, with more training data and advanced computing, the output benefited and rendered the previous detection scheme useless. In the following section, we study the key parameters required for a reliable detector to establish an authentication system for digital media.

## 3. Detection techniques against multimedia attacks

Countering forgery attacks led to the development of detection techniques relying on artifacts related to the in-camera processing module or the post-processing methods. The prior knowledge of the source of the media recordings has been an advantage in detecting forgery; however, without that knowledge, some techniques depend on the artifacts introduced by forgery itself. Techniques based on blind techniques, prior knowledge, and forgery artifacts using the conventional methods are first discussed, followed by neural networks trained to identify the forgery.

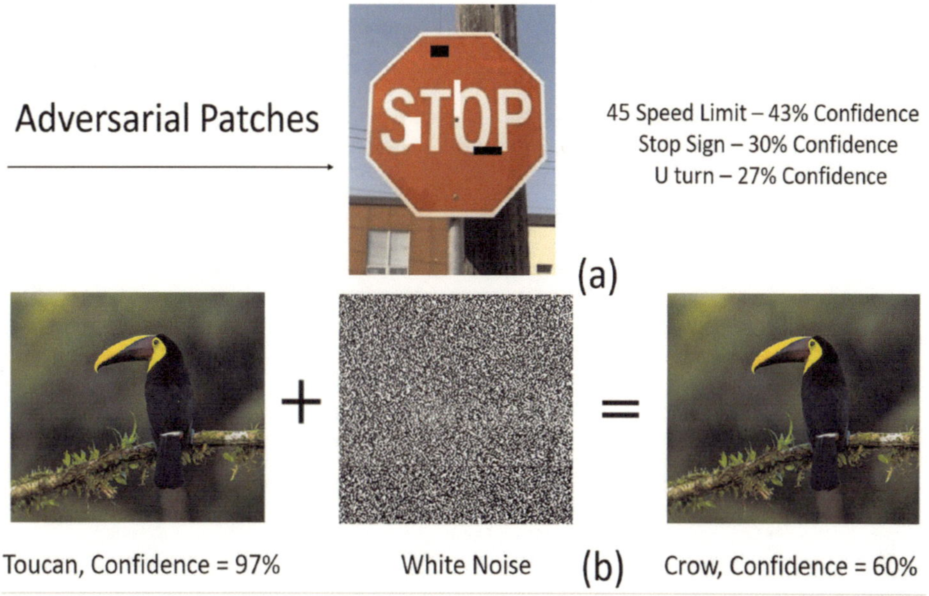


**Figure 3.**
*a) Adversarial patches cause the classifier to wrongly classify the stop sign. b) FGSM attack based on introducing pixel-based perturbations.*

### 3.1 Conventional detection methods

The processing modules present in-camera and post-processing of the media captured result in generating unique features and artifacts, which are exploited to identify frame forgeries. Each image capturing device is equipped with wide or telescopic lenses, where the unique interaction between the lens and the imaging sensor creates chromatic aberrations. A profile of unique chromatic aberrations is created to identify foreign frames inserted from a different lens and sensor [44, 45]. Along with lens distortion artifacts, another module present in in-camera processing after image acquisition is the Color Filter Array (CFA). The CFA is used to record light at a certain wavelength, and the demosaicing algorithm is used to interpolate the missing colors. A periodic pattern emerges due to the in-built CFA module, and whenever a frame is forged, it disrupts the periodic pattern. For frame region splicing attacks, the interrupted periodic pattern from CFA is analyzed to detect the forgery and localize the attack [46, 47].

Each camera sensor manufactured has a unique interaction with the light capturing mechanism due to its sensitivity and photodiode. A unique Sensor Pattern Noise (SPN) is generated for every source camera [48]. It can identify the image acquisition device based on prior knowledge of the camera's sensor noise fingerprint. The SPN noise is similar for RGB and Infrared video; however, it is weaker in Infrared due to low light [49]. Since SPN is used for source device identification, frames moved from an external camera can be identified with any localized in-frame manipulation. The frame and audio acquisition process introduce noise level to the media recordings based on the sensor light sensitivity and localized room reverberations. Using the Error Level Analysis, rich features can be extracted from the noise level present and reveal possible anomalies from image splicing [50].

| Attack type | Attack surface[a] | Trigger | Attack vector[a] | Complexity[b] |
|---|---|---|---|---|
| Frame Manipulation | Visual layer attack (duplication/ shuffling) | • QR code scan<br>• Face/Object Detection<br>• Remote Trigger | • Denial of Service<br>• Malicious Firmware Injection<br>• Live Event Monitoring | • Access: Easy<br>• Low Computation |
| Audio Masking | Auditory layer attack (noise addition/ audio suppression) | • Voice Command<br>• Programmable noise input | • Compromised AED system<br>• Malicious Firmware Injection | • Access: Easy<br>• Low Computation |
| DeepFake Manipulation | • Visual layer attack<br>• Auditory layer attack | • Target Face Detection<br>• Remote Trigger | • Face/Object Detection<br>• Live Event Monitoring<br>• Identity Spoofing | • Access: Medium<br>• High Computation |
| Adversarial Perturbations | • Visual layer attack<br>• Auditory layer attack | • Target Object Detection<br>• Pretrained Noise broadcast | • Object Detection/ Classification<br>• AED Systems | • Access: High (Reconnaissance required)<br>• High Computation |

[a]*Targeted/Compromised systems and attack technique.*
[b]*Attack launching complexity—varied based on ease of access and computational requirements.*

**Table 1.**
*Summary of attack vectors and affected modules.*

In the media capturing post-processing, each compression algorithm uses unique encoding. Therefore, multiple processing of the same media and multiple compression can result in some artifacts identifying prior changes. Analyzing the compression algorithms used by H.264 coding, the presence of any recompression artifacts is used to identify frame manipulations [51]. The spatial and temporal correlation is used to create motion vector features [30, 52]. The de-synchronization caused by removing a group of frames introduces spikes in the Fourier transform of the motion vectors. However, these techniques are sensitive to resolution and noise in the recordings.

The frame manipulations have also inadvertently introduced their unique artifact, and attacks can be identified with prior knowledge of attack nature. Many types of research were developed using custom hand-crafted features. The scale-invariant feature transform key points are used as features for the comparison of duplicated frames in a video recording [53]. The features comprise illumination, noise, rotation, scaling, and small changes in viewpoint. For a continuous frame capture, the standard deviation of residual frames can result in inter-frame duplication detection [54, 55]. Histograms of Oriented Gradients (HOGs) are a unique presentation of pixel value fluctuations, which can be used to identify copy-move forgery based on the HOG feature fluctuation [56]. The optical flow represents the pattern of apparent motion of an image between consecutive frames and its displacement. Using the feature vector designed from the optical flow, copy-move forgery can be identified [31]. Features are generated for each frame and then lexicographically sorted [57]. The Root Mean Square Error (RMSE) is calculated for the frames, and any frame that crosses the threshold is identified as the duplicated frame. However, the technique takes higher processing time due to the sorting and RMSE algorithm and is not applicable in real-time applications.

## 3.2 Machine learning-based detection methods

The development of AI in computer vision has efficiently enabled media processing for forgery detection using trained neural networks. The anomalies introduced in the media recordings result in the forgery-specific artifact, which many research approaches exploit.

### *3.2.1 Artifacts and feature-based detection*

Convolutional neural network (CNN) is the most commonly used frame processing feed-forwarding neural network model, enabling pixel data processing. Forgery attacks such as frame manipulation in the temporal and spatial domain and the DeepFake create an underlying artifact extracted to identify the forgery [58]. In the initial stages of DeepFake development, the resulting media generated visible frame-level artifacts such as inconsistent eye-blinking, face warping, and head-poses. Later, a CNN model is trained to identify the abnormalities introduced by DeepFakes by observing for face warping artifacts [59]. The synthesized face region is spliced into the original image, and a 3D head pose estimation model is created to identify the pose inconsistencies [18]. With the help of pixel information obtained from videos, filters can be designed to identify any tampering. Filters based on discrete cosine transform and video re-quantization errors combined with Deep CNN are used [60].

The DeepFake generation tools are integrated with online conferencing tools to create a fake virtual presence by mimicking a targeted person. The video chat liveness detection in [61] can identify the fake personality due to its fake behavior. The model is trained on behavioral expression in online presence, and any abnormality is marked as fake. For offline media, the audio and video are manipulated to create a video statement; however, the underlying synchronization error for the video lip sync and its corresponding audio are used to identify fake media [62]. To counter DeepFake videos in edge-based computers and online social media, lightweight machine learning models are trained based on the facial presence and its respective spatial and temporal features [63]. Video conferencing solutions are also protected by analyzing the live video stream and passing it through a 3D convolution neural network to predict video segment-wise fakeness scores. The fake online person is identified by the CNN trained on large DeepFake datasets such as Deeperforensics, DFDC, and VoxCeleb.

Along with video forgeries, audio forgeries targeting the AED system in IoT devices like Echo dot by Amazon and Nest Hub by Google are designed. Using the audio perturbations, the AED system misclassifies the incoming voice commands or completely ignores the commands [64]. Training a CNN and recurrent neural network (RNN) [26] has secured the AED system from white noise to disrupt the commands.

### *3.2.2 Fingerprint-based detection*

Modern DeepFake videos are almost perfect without any visual inconsistencies. However, the underlying pixel information is modified due to the project of foreign information on existing media. With advancing DeepFake technology, the current research has developed techniques to identify the underlying pixel fluctuations and use unique fingerprints due to GAN models and in-camera processing. Authors in [65, 66] have identified that GAN leaves unique fingerprints in the media generated from its network. By creating a profile of these unique fingerprints, the forgery can be detected, and the source GAN model used to create the forgery can also be identified. The DeepFake models introduce pixel-level frequency fluctuations, which result in spectral inconsistencies. Inspecting the spectral inconsistencies in a fake image shows that due to the up-sampling convolution of a CNN model in GAN, the frequency artifact is introduced [67, 67]. A filter-based design is used in [68] to highlight the frequency component artifacts introduced by GAN. The two filters used are used in the high-frequency region of an image and the pixel level to observe the changes in the pixels in the background of the image. A biological signature is created from the portrait videos by collecting the signals from different image sections such as facial regions and under image distortions [69].

### *3.2.3 Adversarial training-based detection*

Deep neural networks have been proven to be effective tools in extracting features exclusive to DeepFaked images and can thus detect DeepFake-based image forgery. The traditional approach uses a dataset containing real and fake images to train a CNN model, and to identify artifacts that point to forgery. However, this could lead to the problem of generalization as the validation dataset is often a subset of the training dataset. To avoid this, the images can be preprocessed by using Gaussian Blur and Gaussian noise [70]. Doing so suppresses noise due to pixel-level-high-frequency artifacts. Hybrid models have also been proposed that use multiple streams in parallel to detect fake images [71]. It uses one branch to prepare a model trained on the GoogleNet dataset to differentiate between benign and faked images, and another branch that uses a steganalysis feature extractor to capture low-level details. Results from both the branches are then fused together to formulate the ultimate decision on whether a particular image has been tampered with or not.

There are various approaches to detecting fake or tampered videos using machine learning techniques and can be broadly categorized into those that use biological features for detection, and those that observe spatial and temporal relationships to achieve the same objective. A study proposed a novel approach based on eye blinking to detect tampered videos [72]. It is common knowledge that forgery techniques such as DeepFakes produce little-to-no eye blinking in the fake videos that they produce. Using a combination of CNNs and RNNs that were trained on an eye blinking-based dataset, a binary classifier can be produced, which in turn can be used to detect fake videos with reasonable accuracy. Facial regions of interest were used to train models to differentiate between real and DeepFaked videos [73]. Specifically, photoplethysmography (PPG), which uses color intensities to detect heartbeat variations, was used to train a GAN to distinguish between real and fake face videos. However, the drawback lies in the fact that this method is limited to high-resolution videos containing faces only.

Spatiotemporal analysis-based methods treat videos as a collection of frames related to time. Here, in addition to CNNs, Long-Short Term Memory (LSTM) models are used due to their ability to learn temporal characteristics. One such combination that used a CNN to extract frame level features and an LSTM for temporal sequence analysis was proposed [74]. Simply put, the input to the LSTM is a concatenation of features extracted per frame by the CNN. The final output is a binary prediction as to whether the video is genuine or not. GANs have also been proposed as means of analyzing spatiotemporal relationships of videos. An information theory-based approach was used to study the statistical distribution of fake and real frames, and the differential between them was used to make a decision [75].

## 4. Measure of effective detection techniques

Evaluating the state of the current media authentication system, the existing state-of-the-art technique relies on a fundamental forgery-related artifact or training a deep neural network to identify specific forgery. However, the same deep learning technology has allowed the perpetrator to hijack the existing detection scheme and counteract its purpose. A source device identification methodology used to locate the device used to capture a certain media recording by leveraging the Sensor Pattern Noise fingerprint can be spoofed. The counter method uses a GAN-based approach to inject camera traces into synthetic images, deceiving the detectors into realizing that the synthetic images are real [76]. Development in GAN technology and abundantly available computing resources have generated many fake media that are indistinguishable. A style transfer technique can project facial features into a targeted person and re-create a realistic image [19].

Modern infrastructure relying on machine learning algorithms for seamless people detection and tracking are targeted by adversarial training. A wearable patch can be trained and used to escape the detection or fool the detector into misclassifying the object [29]. The remote trigger mechanism for frame-level attacks is triggered using visual cues and avoids detection by face blur or frame duplication [22]. Tools with simple instructions are designed to allow users to create DeepFake in online video conferences by portraying a targeted person [77].

The need for secure media authentication that spans multiple media categories becomes more and more compelling because of an increase in counterattacks on existing detection techniques. Based on our analysis of the current state-of-the-art detection methods and their counterattacks, here we highlight the key ingredients of the most successful and reliable approaches:

- *Spatial and temporal correlation*: Forgeries involve manipulating spatial frame regions or shuffling the frame itself, which affects the temporal region. A reliable detector should exploit both spatial and temporal correlations to identify forgeries in both layers.

- *Unique Fingerprint*: Deep learning has enabled architectures that are capable of replicating unique device-related fingerprints given sufficient training data. The detector should utilize a fingerprint that is independent of external factors and the device to avoid predictions and re-creation of a unique fingerprint. Inability to control the source of fingerprint generation correlates with difficulty in recreating its unique nature.

- *Multimedia Applicability*: Detectors target specific attacks, which allows a perpetrator to adjust the artifacts and bypass the detection. Both audio and video recordings are the primary input sources for edge devices, and it is equally important to secure both media channels against attacks. A detector should equally account for changes and manipulations in both channels, thereby creating a redundant system capable of dual authentication.

- *Heterogeneous Platform*: Modern smart infrastructure consists of many different types of edge-based IoT smart devices. Each device has its designated functionality relying on either video or audio sensors. Each edge device is also limited in its computational capability due to its power source preservation.

The forgery detection technique should account for enabling its authentication measures across all devices capable of capturing any multimedia.

- *Online Detection*: Attacks are focused on interrupting the active state of the detection system, and most existing techniques are offline systems. Given the state of infrastructure security, it is crucial to immediately raise the alarm upon forgery detection. Enabling instant, online detection can actively observe the media capture and process for any manipulations.

- *Attack Localization*: Lastly, it is important to localize the forgery for further inspection along with attack detection. A detection method that is capable of tracking spatial and temporal changes to the media can locate changes made to the collected samples.

Analyzing the critical traits of a reliable detection system, we propose an environmental fingerprint capable of justifying the qualities aforesaid using the power system frequency. The following section discusses the rationale behind our fingerprint-based authentication system for edge-based IoT devices.

## 5. Environmental fingerprint-based detection

Electrical Network Frequency (ENF) is a power system frequency with a nominal value of 60 Hz in the United States and 50 Hz in most European and Asian countries. The power system frequency fluctuates around its nominal values, making it a time variant, and the resulting signal is referred to as the ENF signal. The ENF-based media authentication was first introduced for audio forgery detection in law enforcement systems [78]. The fluctuations in ENF are similar to a power grid interconnect and originate from the power supply demand, making the fluctuations unique, random, and unpredictable. For audio recordings, ENF is induced in the recordings through electromagnetic induction from being connected to the power grid [78]. Later, it was discovered that battery-operated devices could also capture ENF fluctuations due to the background hum generated by grid-powered devices [79]. In the case of video recordings, ENF is captured in the form of illumination frequency from artificially powered light sources [80]. The capturing of ENF signal through photos depends on the type of imaging sensor used in the camera. For a CCD sensor with a global shutter mechanism, one sample is captured per frame since the whole sensor is exposed at one time instant. However, for a CMOS sensor with a rolling shutter mechanism, each row in the sensor is exposed sequentially, resulting in collecting the ENF samples from spatial and temporal regions of a frame [81, 82].

ENF estimation from media recordings allows many applications due to its time-varying unique nature. For geographical tagging of media recordings, the ENF signal estimated is compared with the global reference database, and its recording location can be identified [83]. Similar fluctuations in ENF signal throughout the power grid are used to synchronize the multimedia recordings in audio and video channels [84]. The fluctuations in ENF and the standard deviations of the signal from its nominal value are observed to study the load effects on the grid and predict blackouts [85].

The estimation of ENF from media recordings is thoroughly studied for a reliable signal estimation [86, 87] and the factors that affect its embedding process [82, 88]. An ENF-based authentication system is integrated for false frame forgery detection in both spatial and temporal regions due to the nature of the ENF signal. In DeFake [77, 89], the distributed nature of ENF is exploited by utilizing ENF as a consensus mechanism for distributed authentication among the edge-based system. The media collected from online systems are processed, and the ENF signal is estimated along with the consensus ground truth signal. With the help of the correlation coefficient, any mismatch in the signal is located, and an alarm is raised. For detailed system implementation and ENF integration techniques, interested readers are referred to papers on the ENF-based authentication system [90, 91].

## 6. State of multimedia authentication

The state of the detection system and forgery attacks never reach an equilibrium where the presented detection scheme can function as a solution for all types of attacks. This chapter discussed the evolution of forgery attacks from subtle frame-level modifications to advanced generated images with fake people, along with its parallel development in detection methods. Based on the critical observations discussed in Section 4, **Table 2** presents a comparison of several current forgery detection techniques.

ENF is a reliable detection method given the signal embedded in the media recordings. The current limitation of this approach involves the recording environment where the ENF-inducing equipment is not present. Due to the absence of artificial lights for outdoor recording, the ENF is not captured in the video recordings. However, in the case of outdoor surveillance recordings, the device is connected to the power grid directly, and the ENF signal is induced in the audio recordings.

Most of the DeepFake detection techniques presented utilize higher computational resources for each frame analysis, and in general, edge devices are not equipped with such power. A different approach would be to design lightweight algorithms utilizing the artifacts or fingerprints for its detection. However, the DeFake approach avoids any training step, and the ENF estimation can be performed in low-computing hardware like Raspberry Pi [91]. Although computer vision has advanced with the emergence of deep learning architecture, DeFake is an environmental fingerprint-based approach relying on signal processing technologies and with encouraging results.

| System | FakeCatcher [69] | FakeBuster [92] | Noiseprint [66] | UpConv [93] | MesoNet [94] | DeFake[a] [77] |
|---|---|---|---|---|---|---|
| Spatial | ✓ | ✓ | ✓ | ✓ | ✓ | ✓ |
| Temporal | ✓ | | | | | ✓ |
| Unique | ✓ | | ✓ | | | ✓ |
| Multimedia | | | | | | ✓ |
| Heterogeneous | | ✓ | | ✓ | | ✓ |
| Online | | ✓ | | ✓ | ✓ | ✓ |
| Localization | ✓ | | ✓ | | | ✓ |

[a]*ENF-based authentication System.*

**Table 2.**
*A comparison of recently proposed forgery detection techniques.*

## 7. Conclusions

The development of forgery attacks has exponentially accelerated with growing computer vision technologies, and the need for a reliable and secure authentication system becomes more compelling. Most detection systems are exploited for their weakness, and attackers frequently launch attacks targeting the system and its security system. This chapter studied the evolution of multimedia attacks using traditional frame-level modification and advanced machine learning-based techniques like DeepFakes. Countering each forgery, we analyzed the detection techniques proposed over time and their progress with the attacks. For a reliable detection and authentication system, we constitute vital ingredients that a system should possess to counter forgery attacks. A thorough analysis and comparison of existing detection techniques are performed to understand the current state of multimedia authentication. Based on the key qualities introduced for a reliable system, we highlight DeFake, an environmental fingerprint-based authentication system, and describe its applications for frame forgeries like a DeepFake attack. Given the state of current edge computing technologies and the constant attacks targeted to disable the system, DeFake is the potential to provide a unique approach for detecting such forgery attacks and protecting the information integrity.

## Acknowledgements

This work is supported by the U.S. National Science Foundation (NSF) via grant CNS-2039342 and the U.S. Air Force Office of Scientific Research (AFOSR) Dynamic Data and Information Processing Program (DDIP) via grant FA9550-21-1-0229. The views and conclusions contained herein are those of the authors and should not be interpreted as necessarily representing the official policies or endorsements, either expressed or implied, of the U.S. Air Force.

## Abbreviations

| | |
|---|---|
| ML | Machine Learning |
| AI | Artificial Intelligence |
| IoT | Internet of Things |
| VoIP | Voice over Internet Protocol |
| DL | Deep Learning |
| GAN | General Adversarial Networks |
| AED | Audio Event Detector |
| PTZ | Pan-Tilt-Zoom |
| CFA | Color Filter Array |
| SPN | Sensor Pattern Noise |
| HOG | Histogram of Oriented Gradients |
| RMSE | Root Mean Square Error |
| CNN | Convolutional Neural Network |
| RNN | Recurrent Neural Network |
| LSTM | Long Short Term Memory |
| PPG | Photoplethysmography |

| | |
|---|---|
| ENF | Electrical Network Frequency |
| CMOS | Complementary Metal Oxide Semiconductor |
| CCD | Charge-Coupled Device |

## Author details

Deeraj Nagothu, Nihal Poredi and Yu Chen*
Department of Electrical and Computer Engineering, Binghamton Univerisity, Binghamton, New York, USA

*Address all correspondence to: ychen@binghamton.edu

## References

[1] Chen J, Li K, Deng Q, Li K, Philip SY. Distributed Deep Learning Model for Intelligent Video Surveillance Systems with Edge Computing. NY, United States: IEEE Transactions on Industrial Informatics; 2019

[2] Nikouei SY, Chen Y, Song S, Choi BY, Faughnan TR. Toward intelligent surveillance as an edge network service (isense) using lightweight detection and tracking algorithms. IEEE Transactions on Services Computing. 2019;**14**(6): 1624-1637

[3] Obermaier J, Hutle M. Analyzing the security and privacy of cloud-based video surveillance systems. In: Proc. 2nd ACM Int. Work. IoT Privacy, Trust. Secur. NY, United States: ACM; 2016. pp. 22-28

[4] Shi W, Cao J, Zhang Q, Li Y, Xu L. Edge computing: Vision and challenges. IEEE Internet of Things Journal. 2016; **3**(5):637-646

[5] Chen N, Chen Y, Blasch E, Ling H, You Y, Ye X. Enabling smart urban surveillance at the edge. In: 2017 IEEE International Conference on Smart Cloud (Smart Cloud). NY, United States: IEEE; 2017. pp. 109-119

[6] Chen N, Chen Y, Ye X, Ling H, Song S, Huang CT. Smart city surveillance in fog computing. In: Advances in Mobile Cloud Computing and Big Data in the 5G Era. Cham, Switzerland: Springer; 2017. pp. 203-226

[7] Nikouei SY, Xu R, Nagothu D, Chen Y, Aved A, Blasch E. Real-time index authentication for event-oriented surveillance video query using blockchain. In: 2018 IEEE Int. Smart Cities Conf. ISC2 2018. 2019

[8] Xu R, Nagothu D, Chen Y. Decentralized video input authentication as an edge Service for Smart Cities. IEEE Consumer Electronics Magazine. 2021; **10**(6):76-82

[9] Mowery K, Wustrow E, Wypych T, Singleton C, Comfort C, Rescorla E, et al. Security analysis of a full-body scanner. In: 23rd USENIX Security Symposium (USENIX Security 14). San Diego, CA, United States; 2014. pp. 369-384

[10] Olsen M. Beware, Even Things on Amazon Come with Embedded Malware. 2016. Available from: https://artfulhacker.com/post/142519805054/beware-even-things-on-amazon-come

[11] Costin A. Security of Cctv and video surveillance systems: Threats, vulnerabilities, attacks, and mitigations. In: Proc. 6th Int. Work. Trust. Embed. Devices. NY, United States: ACM; 2016. pp. 45-54

[12] Carlini N, Mishra P, Vaidya T, Zhang Y, Sherr M, Shields C, et al. Hidden voice commands. In: 25th USENIX Security Symposium (USENIX Security 16). Austin, TX, United States; 2016. pp. 513-530

[13] Goodfellow I, Pouget-Abadie J, Mirza M, Xu B, Warde-Farley D, Ozair S, et al. Generative Adversarial Nets. In: Advances in Neural Information Processing Systems. Montreal, Canada; 2014

[14] Verdoliva L. Media forensics and deepfakes: an overview. IEEE Journal of Selected Topics in Signal Processing. NY, United States. 2020;**14**(5):910-32

[15] Westerlund M. The emergence of Deepfake technology: A review. Technology Innovation and Management Review. 2019;**9**(11):40-53

[16] Nagothu D, Chen YY, Blasch E, Aved A, Zhu S. Detecting malicious false frame injection attacks on surveillance Systems at the Edge Using Electrical Network Frequency Signals. Sensors (Basel). 2019;**19**(11):1-19

[17] Wolfgang RB, Delp EJ. A watermark for digital images. In: Proceedings of 3rd IEEE International Conference on Image Processing. Lausanne, Switzerland: IEEE; 1996. pp. 219-222

[18] Yang X, Li Y, Lyu S. Exposing deepfakes using inconsistent head poses. In: ICASSP 2019-2019 IEEE International Conference on Acoustics, Speech and Signal Processing (ICASSP). Brighton, United Kingdom: IEEE; 2019. pp. 8261-8265

[19] Karras T, Laine S, Aila T. A style-based generator architecture for generative adversarial networks. In: Proceedings of the IEEE/CVF conference on computer vision and pattern recognition. Long beach, CA, United States; 2019. pp. 4401-4410

[20] Mena DM, Papapanagiotou I, Yang B. Internet of things: Survey on security. Information Security Journal: A Global Perspective. Philadelphia, PA, United Stated. 2018;**27**(3):162-82. DOI: 10.1080/19393555.2018.1458258

[21] Kharraz A, Kirda E, Robertson W, Balzarotti D, Francillon A. Optical delusions: A study of malicious QR codes in the wild. In: 2014 44th Annual IEEE/IFIP International Conference on Dependable Systems and Networks. Atlanta, GA, United States; 2014. pp. 192-203

[22] Nagothu D, Schwell J, Chen Y, Blasch E, Zhu S. A study on smart online frame forging attacks against video surveillance system. In: Proc. SPIE - Int. Soc. Opt. Eng. Bellingham, Washington, United States; 2019

[23] Zhang C, Shahriar H, Riad ABMK. Security and privacy analysis of wearable health device. In: 2020 IEEE 44th Annual Computers, Software, and Applications Conference (COMPSAC). Los Alamitos, CA, United States; 2020. pp. 1767-1772

[24] Kriesel D. Xerox Scanners/Photocopiers Randomly Alter Numbers in Scanned Documents. 2017. https://www.dkriesel.com/en/blog/ 2013/0802_xerox-workcentres_are_switching_written_numbers_when_scanning

[25] Stamm MC, Lin WS, Liu KR. Temporal forensics and anti-forensics for motion compensated video. IEEE Transactions on Information Forensics and Security. 2012;**7**(4):1315-1329

[26] dos Santos R, Kassetty A, Nilizadeh S. Disrupting audio event Detection deep neural networks with white noise. Technologies. 2021;**9**(3):64

[27] Akhtar N, Mian A. Threat of adversarial attacks on deep learning in computer vision: A survey. IEEE Access. 2018;**6**:14410-14430

[28] Quan W, Nagothu D, Poredi N, Chen Y. Cri PI: An efficient critical pixels identification algorithm for fast one-pixel attacks. In: Sensors and Systems for Space Applications. Bellingham, Washington, United States: SPIE; 2021. pp. 83-99

[29] Thys S, Van Ranst W, Goedeme T. Fooling automated surveillance cameras: Adversarial patches to attack person Detection. In: Proceedings of the IEEE/CVF Conference on Computer Vision and Pattern Recognition Workshops. Long Beach, CA, United States; 2019

[30] Wang W, Farid H. Exposing digital forgeries in video by detecting

duplication. In: Proc. 9th Work. Multimed. Secur. Dallas, Texas, United States: ACM; 2007. pp. 35-42

[31] Al-Sanjary OI, Abdullah Ahmed A, Ahmad HB, Ali MAM, Mohammed MN, Irsyad Abdullah M, et al. Deleting object in video copy-move forgery Detection based on optical flow concept. In: 2018 IEEE Conference on Systems, Process and Control (ICSPC). Melaka, Malaysia; 2018. pp. 33-38

[32] Ulutas G, Ustubioglu B, Ulutas M, Nabiyev V. Frame duplication/mirroring Detection method with binary features. IET Image Processing. 2017;**11**(5): 333-342

[33] Su L, Li C. A novel passive forgery Detection algorithm for video region duplication. Multidimensional Systems and Signal Processing. 2018;**29**(3): 1173-1190

[34] Wahab AWA, Bagiwa MA, Idris MYI, Khan S, Razak Z, Ariffin MRK. Passive video forgery detection techniques: A survey. In: 2014 10th Int. Conf. Inf. Assur. Al Ain, United Arab Emirates; 2014. pp. 29-34

[35] Korshunov P, Marcel S. Deep fakes: a new threat to face recognition? Assessment and detection. In: 2018 Computer Vision and Pattern Recognition. Salt Lake City, Utah, United States; 2018. DOI: 10.48550/arXiv.1812.08685

[36] Bitouk D, Kumar N, Dhillon S, Belhumeur P, Nayar SK. Face swapping: Automatically replacing faces in photographs. In: ACM SIGGRAPH 2008 Papers. SIGGRAPH '08. New York, NY, USA: Association for Computing Machinery; 2008. pp. 1-8

[37] Perov I, Gao D, Chervoniy N, Liu K, Marangonda S, Umé C, et al. Deep face lab: Integrated, flexible and extensible face-swapping framework. In: 2021 Computer Vision and Pattern Recognition. NY, United States; 2021

[38] Thies J, Zollhofer M, Stamminger M, Theobalt C, Niessner M. Face 2Face: Real-time face capture and Reenactment of RGB videos. In: Proceedings of the IEEE Conference on Computer Vision and Pattern Recognition. Las Vegas, Nevada; 2016. pp. 2387-2395

[39] Chan C, Ginosar S, Zhou T, Efros AA. Everybody dance now. In: Proceedings of the IEEE/CVF International Conference on Computer Vision. Seoul, South Korea; 2019. pp. 5933-5942

[40] Eykholt K, Evtimov I, Fernandes E, Li B, Rahmati A, Xiao C, et al. Robust physical-world attacks on deep learning visual classification. In: Proceedings of the IEEE Conference on Computer Vision and Pattern Recognition. Salt Lake City, Utah; 2018. pp. 1625-1634

[41] Goodfellow IJ, Shlens J, Szegedy C. Explaining and harnessing adversarial examples. In: 6th International Conference on Learning Representations (ICLR). San Diego, CA, United States; 2015. DOI: 10.48550/arXiv.1412.6572

[42] Papernot N, McDaniel P, Jha S, Fredrikson M, Celik ZB, Swami A. The limitations of deep learning in adversarial settings. In: IEEE European Symposium on Security and Privacy (Euro S and P). Saarbrücken, Germany: IEEE; 2016. pp. 372-387

[43] Descript | Create Podcasts, Videos, and Transcripts. 2021. Available from: https://www.descript.com/

[44] Yerushalmy I, Hel-Or H. Digital image forgery detection based on lens and sensor aberration. International

Journal of Computer Vision. 2011;**92**(1): 71-91

[45] Fu H, Cao X. Forgery authentication in extreme wide-angle Lens using distortion Cue and fake saliency map. IEEE Transactions on Information Forensics and Security. 2012;**7**(4): 1301-1314

[46] Bayram S, Sencar H, Memon N, Avcibas I. Source camera identification based on CFA interpolation. In: IEEE International Conference on Image Processing. Genoa, Italy; 2005

[47] Cao H, Kot AC. Accurate Detection of Demosaicing regularity for digital image forensics. IEEE Transactions on Information Forensics and Security. 2009;**4**(4):899-910

[48] Lukas J, Fridrich J, Goljan M. Digital camera identification from sensor pattern noise. IEEE Transactions on Information Forensics and Security. 2006;**1**(2):205-214

[49] Hyun DK, Lee MJ, Ryu SJ, Lee HY, Lee HK. Forgery detection for surveillance video. In: Jin JS, Xu C, Xu M, editors. The Era of Interactive Media. New York, NY: Springer; 2013. pp. 25-36

[50] Cozzolino D, Poggi G, Verdoliva L. Splicebuster: A new blind image splicing detector. In: 2015 IEEE International Workshop on Information Forensics and Security (WIFS). Rome, Italy; 2015. pp. 1-6

[51] González Fernández E, Sandoval Orozco AL, García Villalba LJ. Digital video manipulation Detection technique based on compression algorithms. IEEE Transactions on Intelligent Transportation Systems. 2022;**23**(3): 2596-2605

[52] Wang W, Farid H. Exposing digital forgeries in interlaced and deinterlaced video. IEEE Transactions on Information Forensics and Security. 2007;**2**(3): 438-449

[53] Kharat J, Chougule S. A passive blind forgery Detection technique to identify frame duplication attack. Multimedia Tools and Applications. 2020;**79**(11): 8107-8123

[54] Fadl SM, Han Q, Li Q. Authentication of surveillance videos: Detecting frame duplication based on residual frame. Journal of Forensic Sciences. 2018;**63**(4): 1099-1109

[55] Bestagini P, Milani S, Tagliasacchi M, Tubaro S. Local tampering Detection in video sequences. In: 2013 IEEE 15th International Workshop on Multimedia Signal Processing (MMSP). Pula, Sardinia, Italy; 2013. pp. 488-493

[56] Subramanyam AV, Emmanuel S. Video forgery Detection using HOG features and compression properties. In: 2012 IEEE 14th International Workshop on Multimedia Signal Processing (MMSP). 2012. pp. 89-94

[57] Singh VK, Pant P, Tripathi RC. Detection of frame duplication type of forgery in digital video using sub-block based features. Int. Conf. Digit. Forensics Cyber Crime. Seoul, South Korea: Springer; 2015. pp. 29–38

[58] Güera D, Delp EJ. Deepfake video detection using recurrent neural networks. In: 2018 15th IEEE International Conference on Advanced Video and Signal Based Surveillance (AVSS). Auckland, New Zealand; 2018

[59] Li Y, Lyu S. Exposing deepfake videos by detecting face warping

artifacts. 2018. In Computer Vision and Pattern Recognition. Salt Lake City, Utah, United Stated; 2018

[60] Zampoglou M, Markatopoulou F, Mercier G, Touska D, Apostolidis E, Papadopoulos S, et al. Detecting tampered videos with multimedia forensics and deep learning. In: Kompatsiaris I, Huet B, Mezaris V, Gurrin C, Cheng WH, Vrochidis S, editors. Multi Media Modeling. Lecture Notes in Computer Science. Cham: Springer International Publishing; 2019. pp. 374-386

[61] Liu H, Li Z, Xie Y, Jiang R, Wang Y, Guo X, et al. Live Screen: Video Chat Liveness Detection Leveraging Skin Reflection. In: IEEE INFOCOM 2020 - IEEE Conference on Computer Communications. Toronto, ON, Canada: IEEE; 2020. pp. 1083-1092

[62] Zhou Y, Lim SN. Joint audio-visual Deepfake Detection. In: Proceedings of the IEEE/CVF International Conference on Computer Vision. Montreal, Canada; 2021. pp. 14800-14809

[63] Chen HS, Rouhsedaghat M, Ghani H, Hu S, You S, CCJ K. Defake Hop: A Light-Weight High-Performance Deepfake Detector. In: 2021 IEEE International Conference on Multimedia and Expo (ICME). Shenzhen, China; 2021

[64] Santos RD, Nilizadeh S. Audio attacks and defenses against AED systems – A practical study. In: Audio and Speech Processing. Ithaca, NY, United States; 2021. DOI: 10.48550/arXiv.2106.07428

[65] Marra F, Gragnaniello D, Verdoliva L, Poggi G. Do GANs Leave Artificial Fingerprints? In: 2019 IEEE Conference on Multimedia Information Processing and Retrieval (MIPR). San Jose, CA, United States; 2019. pp. 506-511

[66] Cozzolino D, Verdoliva L. Noiseprint: A CNN-based camera model fingerprint. IEEE Transactions on Information Forensics and Security. 2020;**15**:144-159

[67] Durall R, Keuper M, Pfreundt FJ, Keuper J. Unmasking deep fakes with simple features. In: Computer Vision and Pattern Recognition. Seattle, Washington, United States; 2020

[68] Jeong Y, Kim D, Min S, Joe S, Gwon Y, Choi J. BiHPF: Bilateral high pass filters for robust deepfake detection. In: Proceedings of the IEEE/CVF Winter Conference on Applications of Computer Vision. Waikoloa, HI, United States; 2022. pp. 48-57

[69] Ciftci UA, Demir I, Yin L. Fake catcher: Detection of synthetic portrait videos using biological signals. IEEE Transactions on Pattern Analysis and Machine Intelligence. 2020; **2020**:1-1

[70] Xuan X, Peng B, Wang W, Dong J. On the generalization of GAN image forensics. In: Chinese Conference on Biometric Recognition. Zhuzhou, China: Springer; 2019. pp. 134-141

[71] Zhou P, Han X, Morariu VI, Davis LS. Two-stream neural networks for tampered face detection. In: 2017 IEEE Conference on Computer Vision and Pattern Recognition Workshops (CVPRW). Honolulu, Hawaii, United States: IEEE; pp. 1831-1839

[72] Li Y, Chang MC, Lyu S. In ictu oculi: Exposing ai created fake videos by detecting eye blinking. In: 2018 IEEE International workshop on information

forensics and security (WIFS). Hong Kong, China: IEEE; 2018. pp. 1-7

[73] Ciftci UA, Demir I, Yin L. How do the hearts of deep fakes beat? deep fake source detection via interpreting residuals with biological signals. In: 2020 IEEE International Joint Conference on Biometrics (IJCB). Houston, TX, United States: IEEE; 2020. pp. 1-10

[74] Güera D, Delp EJ. Deepfake video detection using recurrent neural networks. In: 2018 15th IEEE International Conference on Advanced Video and Signal Based Surveillance (AVSS). Auckland, New Zealand: IEEE; 2018. pp. 1-6

[75] Agarwal S, Girdhar N, Raghav H. A novel neural model based framework for detection of GAN generated fake images. In: 2021 11th International Conference on Cloud Computing, Data Science Engineering (Confluence). Uttar Pradesh, India; 2021. pp. 46-51

[76] Cozzolino D, Thies J, Rossler A, Niessner M, Verdoliva L. Spo C: Spoofing camera fingerprints. In: Proceedings of the IEEE/CVF Conference on Computer Vision and Pattern Recognition. NY, United States; 2021. pp. 990-1000

[77] Nagothu D, Xu R, Chen Y, Blasch E, Aved A. DeFake: Decentralized ENF-consensus based deep fake detection in video conferencing. In: IEEE 23rd International Workshop on Multimedia Signal Processing. Tampere, Finland; 2021

[78] Grigoras C. Applications of ENF criterion in forensic audio, video, computer and telecommunication analysis. Forensic Science International. 2007;**167**(2–3):136-145

[79] Chai J, Liu F, Yuan Z, Conners RW, Liu Y. Source of ENF in battery-powered digital recordings. In: Audio Eng. Soc. Conv. Rome, Italy: Audio Engineering Society; 2013

[80] Garg R, Varna AL, Hajj-Ahmad A, Wu M. "Seeing" ENF: Power-signature-based timestamp for digital multimedia via optical sensing and signal processing. IEEE Transactions on Information Forensics and Security. 2013;**8**(9):1417-1432

[81] Vatansever S, Dirik AE, Memon N. Analysis of rolling shutter effect on ENF based video forensics. IEEE Transactions on Information Forensics and Security. 2019;**14**(7):2262-2275

[82] Nagothu D, Chen Y, Aved A, Blasch E. Authenticating video feeds using electric network frequency estimation at the edge. EAI Endorsed Transactions on Security and Safety. 2021;**7**(24):e4-e4

[83] Wong CW, Hajj-Ahmad A, Wu M. Invisible geo-location signature in a single image. In: 2018 IEEE International Conference on Acoustics, Speech and Signal Processing (ICASSP). Calgary, Alberta, Canada: 2018. pp. 1987-1991

[84] Vidyamol K, George E, Jo JP. Exploring electric network frequency for joint audio-visual synchronization and multimedia authentication. In: 2017 International Conference on Intelligent Computing, Instrumentation and Control Technologies (ICICICT). Kannur, Kerala, India; 2017. pp. 240-246

[85] Liu Y, You S, Yao W, Cui Y, Wu L, Zhou D, et al. A distribution level wide area monitoring system for the electric power grid–FNET/grid eye. IEEE Access. 2017;**5**:2329-2338

[86] Hua G, Zhang H. ENF signal enhancement in audio recordings. IEEE Transactions on Information Forensics and Security. 2020;**15**:1868-1878

[87] Hajj-Ahmad A, Garg R, Wu M. Spectrum combining for ENF signal estimation. IEEE Signal Processing Letters. 2013;**20**(9):885-888

[88] Hajj-Ahmad A, Wong CW, Gambino S, Zhu Q, Yu M, Wu M. Factors affecting ENF capture in audio. IEEE Transactions on Information Forensics and Security. 2019;**14**(2): 277-288

[89] Xu R, Nagothu D, Chen Y. Econ ledger: A proof-of-ENF consensus based lightweight distributed ledger for IoVT networks. Future Internet. 2021; **13**(10):248

[90] Nagothu D, Xu R, Chen Y, Blasch E, Aved A. Detecting compromised edge smart cameras using lightweight environmental fingerprint consensus. In: Proceedings of the 19th ACM Conference on Embedded Networked Sensor Systems. New York, NY, USA: Association for Computing Machinery; 2021. pp. 505-510

[91] Nagothu D, Xu R, Chen Y, Blasch E, Aved A. Deterring Deepfake attacks with an electrical network frequency fingerprints approach. Future Internet. 2022;**14**(5):125

[92] Mehta V, Gupta P, Subramanian R, Dhall A. FakeBuster: a DeepFakes detection tool for video conferencing scenarios. In: 26th International Conference on Intelligent User Interfaces-Companion. College Station, TX, United States; 2021. pp. 61-63

[93] Durall R, Keuper M, Keuper J. Watch your up-convolution: CNN based generative deep neural networks are failing to reproduce spectral distributions. In: Proceedings of the IEEE/CVF Conference on Computer Vision and Pattern Recognition. NY, United States; 2020. pp. 7890-7899

[94] Afchar D, Nozick V, Yamagishi J, Echizen I. Meso net: A compact facial video forgery Detection network. In: 2018 IEEE International Workshop on Information Forensics and Security (WIFS). Hong Kong, China; 2018. pp. 1-7

Chapter 5

# Surveillance with UAV Videos

*İbrahim Delibaşoğlu*

## Abstract

Unmanned aerial vehicles (UAVs) and drones are now accessible to everyone and are widely used in civilian and military fields. In military applications, UAVs can be used in border surveillance to detect or track any moving object/target. The challenge of processing UAV images is the unpredictable background motions due to camera movement and small target sizes. In this chapter, a short literature brief will be discussed for moving object detection and long-term object tracking. Publicly available datasets in the literature are introduced. General approaches and success rates in the proposed methods are evaluated and approach to how deep learning-based solutions can be used together with classical methods are discussed. In addition to the methods in the literature for moving object detection problems, possible solution approaches for the challenges are also shared.

**Keywords:** surveillance, moving object, motion detection, foreground detection, object tracking, long-term tracking, UAV video, drones

## 1. Introduction

Unmanned aerial vehicles (UAV) and drones are now accessible to everyone and are widely used in civilian and military fields. Considering security applications, drones could be used in applications such as surveillance, target detection and tracking. Drone surveillance allows us to continuously gather information about a tracked target from a distance. So drones with the capabilities of features such as object tracking, autonomous navigation, and event analysis are a hot topic in computer vision society. The challenge of processing drone videos is the unpredictable background motion due to camera movement. In this chapter, a short literature brief, potential approaches to improve the moving object detection performance, will be discussed and publicly available datasets in the literature will be introduced. In addition, the current situation of deep learning-based solutions, which give good results in many research areas, in motion detection and potential solutions will be discussed. General approaches and success rates in the proposed methods will be shared, and approaches on how deep learning-based solutions can be used together with classical methods will be proposed. In brief, we propose some post-processing techniques to improve the performance of background modeling-based methods, and software architecture to speed up operations by dividing them into small parts.

Section 2 represents moving target detection issues from UAV videos, while Section 2.1 represents how to build a simple background model. Section 2.2 introduces sample datasets for moving target detection and Section 2.3 gives potential approaches to enhance the background modeling approach for moving target detection. Some object tracking methods that can be used together with moving object detection and Convolutional Neural Network (CNN) based methods are emphasized in Sections 3 and 4, respectively. Finally, the conclusion is discussed in Section 5.

## 2. Moving object detection

The problem of detecting moving objects is a computer vision issue that is needed in areas such as real-time object tracking, event analysis and security applications. Based on the computer vision literature carried out in recent years, it is a problem that has been studied extensively [1]. The purpose of moving object detection is to classify the image as foreground and background. The classification could be challenging according to factors such as the motion state of the camera, ambient lighting, background cluttering, and dynamic changes in the background. Images obtained from cameras mounted on drones have a free motion, and it causes much background motion (also called global motion in the literature). Another important issue is that these images could be taken in very different regions such as mountains, nature, forests, cities, rural areas, and they can contain very small targets according to the altitude of the UAV.

In moving object detection applications, the aim is to have high accuracy as well as real-time operation of the application. When the studies carried out in the literature are examined, it is seen that subtraction of consecutive frames, background modeling and optical flow-based methods are used. Although the subtraction of consecutive frames method works fast and can adapt quickly to background changes, the success rate is very low [2]. In the background modeling approach, a background model (an image formed as a result of the average of the previous *n* frames) is extracted using frames history [3]. Classical image processing techniques [4], statistical methods [5–7] and neural networks [8] have been used for background modeling in the literature. Gaussian mixture model (GMM) [9] builds a Gaussian distribution model for each pixel and adaptive GMM [7] improves it for dynamic background. Kim et al. [10] propose a spatio-temporal Gaussian model minimizing image registration errors. Zhong Z. et al. [11] propose a background updating strategy performing at both pixel and object levels and apply a pixel-based adaptive segmentation method. Dual-target non-parametric background modeling method [12] proposes a dual-target updating strategy to eliminate false detection caused by background movements and illumination changes. Scene conditional background update method [13], named SCBU, builds a statistical background model without contamination of the foreground pixels. Background subtraction is applied between current frame and updated background model while calculated foreground likelihood map is used to extract initial foreground regions by applying high and low threshold values. In MCD method [6], a dual-mode single Gaussian model is proposed with age, mean and variance of each pixel, and it compensates for the camera motion by mixing neighboring approaches. Simple threshold with respect to the variance is applied in MCD method for foreground detection. Yu et al. [14] use a candidate background model similar to MCD and they propose a method to update candidate or main background model pixels in each frame. In background subtraction step, they apply the neighborhood subtraction approach, which takes into account the neighbors of each pixel. BSDOF [15] method extracts candidate foreground masks with background subtraction and applies threshold for variance value of each pixel. In background subtraction process, also uses dense optical flow to weigh the difference for each pixel. Then, it obtains a final mask with the combination of candidate masks and region growing strategy using candidate masks. Thus, false detection is largely eliminated.

For background modeling approach in moving cameras (such as cameras mounted to UAVs), global motion is generally eliminated by using homography matrix obtained by Lucas Kanade [16] (KLT) and RANSAC [17] methods. Selected points in the previous frame are tracked in the current frame with KLT and homography matrix representing global (camera) motion is calculated with RANSAC method. Then, previous frame or background model is warped to the current frame to eliminate the global motion. Sample grid-based selected points and estimated positions are visualized as flow vectors in **Figure 1**.

One of the biggest problems in using only pixel intensity values is that these kinds of methods are so sensitive to illumination changes and registration errors caused by homography errors. As a solution to these issues, different features such as texture [18], edge [19] and haar-like [20] are proposed in the literature. Edge and texture features can better address the illumination change issue and also eliminate the ghosting effect left by foreground objects. Local Binary Pattern (LBP) and its variants [21, 22] are other types of texture feature used for foreground detection in the literature. In addition to such additional features, deep learning methods that offer effective solutions to many problems have also been used in the foreground detection problem. For this purpose, FlowNet2 [23] architecture estimating optical flow vectors are used in foreground detection problems [24]. Optical flow means the displacement of each pixel in consecutive frames. KLT method is also an optical flow method that tracks the given points in the consecutive frame and it is categorized as sparse optical flow. On other hand, estimating pixel displacement of each pixel is called dense optical flow. FlowNet2 is one of the most known architectures which also has publicly available pre-trained weights. The disadvantage of deep learning methods is that they require much computational cost, especially for high-dimensional images, and may not perform well for so small targets due to the training image dimensions and

**Figure 1.**
*Visualization of flow vectors for grid points.*

contents. Considering that UAV images may contain a lot of small targets, it can be thought that the optical flow model to be trained with small moving object images could perform better. On the other hand processing, high-dimensional input images require much RAM in the GPU. **Figure 2** shows sample visualization of optical flows for FlowNetCSS (which is pre-trained model that mostly detects the small changes and more lightweight according to FlowNet2), Farneback and Nvidia Optical Flow (NVOF). FlowNetCSS is a sub-network of Flownet2.

In this work, we have used FlowNet pre-trained weights which have been trained on MPI-Sintel dataset [25] containing images with the resolution of 1024 × 436. **Figure 3** shows the FlowNetCSS output on 1920 × 1080 resolution images from PESMOD dataset [26]. In **Figure 4**, the model is runned for a patch of the frames instead of full resolution and it performs better for the small targets (two people hiking in the mountain). Simple thresholding could be applied for optical flow matrices and the foreground mask showing the moving pixels could be obtained directly. But for small targets, it may be useful to process the small regions as shown in **Figure 4**. Global motion compensation with homography matrix may also be used

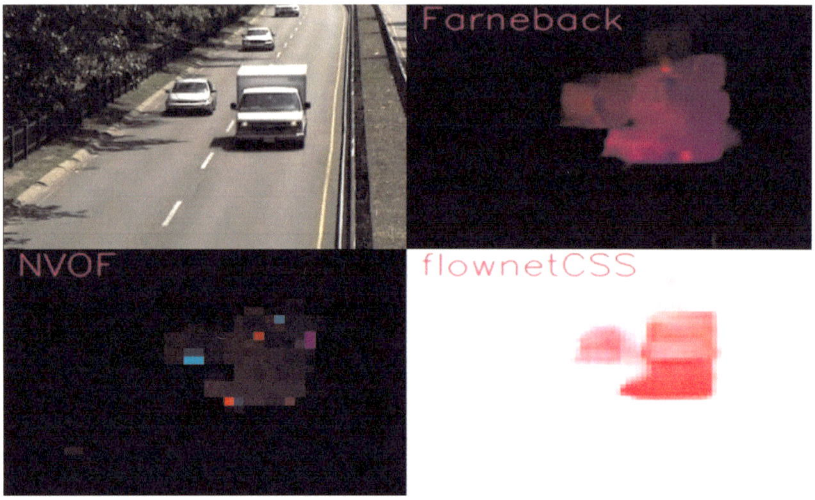


**Figure 2.**
*Visualization of optical flow vectors of FlowNetCSS, Farneback and NVOF.*

**Figure 3.**
*FlowNet visualization on PESMOD [26] sample frames.*

**Figure 4.**
*FlowNet visualization on a patch of PESMOD [26] sample frames.*

**Figure 5.**
*(a) Sample frame (b) Background model μ image.*

before estimating dense optical flow, so simple thresholding can give the moving pixels with better accuracy.

## 2.1 Building a background model

Consider that $H$ represents homography matrix between frames in time $t-1$ and $t$. The background model $B$ at time of *t-1* is warped to the current frame by using Eq. (1). Thus, the pixels in the background and current frame are aligned to handle global motion. $\alpha(i)$ represents the learning rate of each pixel while $\mu_t(i)$ represents average pixel values. The background model $B$ consists of mean and learning values as shown in Eq. (2) and (3).

$$B_t = H_{t-1}B_{t-1} \tag{1}$$

$$\alpha(i) = \frac{1}{age(i)} \tag{2}$$

$$\mu_t(i) = (1 - \alpha_t(i))\mu_{t-1}(i) + \alpha(x)\ \ I_t(i) \tag{3}$$

In the equations, $I$ represents a frame while $i$ represents a pixel in a frame. Learning rates ($\alpha$) is determined with the *age* value of each pixel. Sample frame and background image is shown in the **Figure 5** for maximum age value 30. It is also important to set pixels whose age is less than a fixed threshold value to zero. Because the pixels that have just entered the frame need to wait for a while to be evaluated. After building the background model, current frame is subtracted from $\mu$ image to obtain a foreground mask. But using a simple model with only RGB coloir features is so sensitive to errors like shadow, ghost effect, illumination changes and background motion. Thus, it is important to use extra texture features for background modeling as mentioned in Section 2. In Chapter 2.3 we discuss some approaches to improve the performance of BSDOF method using color features effectively.

## 2.2 Datasets

Changedetection.net (CDNET) [27] dataset is a large-scale video dataset consisting of 11 different categories, but only PTZ subsequence consists of images taken by moving camera. PTZ sequence does not include free motion so it is not so appropriate to evaluate motion detection problem for UAV images. SCBU dataset [13] includes images of walking pedestrians with a free motion camera. The VIVID [28] dataset consisting of aerial images is a good candidate to evaluate moving object detection methods. It consists of moving vehicle images and has a resolution of 640x480. PESMOD [15] dataset represents a new challenging high-resolution dataset for evaluation of small moving object detection methods. It includes eight different sequences with a resolution of 1920x1080 and consists of small moving targets (vehicles and humans). PESMOD dataset contains totally of 4107 frames and 13,834 labeled bounding boxes for moving targets. The details of each sequence is given in **Table 1**.

Average precision ($Pr$), recall ($R$) and f1 ($F_1$) score values of MCD, SCBU and BSDOF methods for PESMOD dataset are given in **Table 2**. In the Eq. (4), $FP$ refers to wrongly detected boxes, $TP$ refers to the number of true detections, and $FN$ refers to ground truth boxes that is missed by the method. $Pr$ indicates the accuracy of positive predictions (estimated as motion) while $R$ (also named sensitivity) represents the ratio of the number of pixels correctly classified as foreground (motion) to the actual

| Sequence name | Number of frames | Number of moving objects |
|---|---|---|
| *Pexels-Elliot-road* | 664 | 3416 |
| *Pexels-Miksanskiy* | 729 | 189 |
| *Pexels-Shuraev-trekking* | 400 | 800 |
| *Pexels-Welton* | 470 | 1129 |
| *Pexels-Marian* | 622 | 2791 |
| *Pexels-Grisha-snow* | 115 | 1150 |
| *Pexels-zaborski* | 582 | 3290 |
| *Pexels-Wolfgang* | 525 | 1069 |

**Table 1.**
*The details of PESMOD dataset.*

| Metrics | MCD [6] | SCBU [13] | BSDOF [15] |
|---|---|---|---|
| Precision | 0.3928 | 0.3248 | **0.4890** |
| Recall | **0.4163** | 0.3127 | 0.4061 |
| F1 score | 0.2856 | 0.3072 | **0.3898** |

*Bold values in the table represents the best score for each row.*

**Table 2.**
*Comparison of average precision, recall and f1 score values of MCD, SCBU and BSDOF methods on PESMOD dataset.*

number of foreground pixels. $F_1$ score is the combination of $Pr$ and $R$, and is equal to 1 for perfect classification

$$Pr = \frac{TP}{TP + FP}, R = \frac{TP}{TP + FN}, F_1 = \frac{2^* Pr^* R}{Pr + R} \quad (4)$$

The BSDOF method is suitable to implement in the GPU. It runs at about 26 fps for 1920x1080 on a PC with Ubuntu 18.04 operation system, AMD Ryzen 53,600 processor with 16 GB RAM, and Nvidia GeForce RTX2070 graphic card. MCD runs at about 8 fps on the same machine. SCBU is also implemented for CPU and we have used the binary files. So that we could not measure the processing time of the SCBU method.

### 2.3 Prospective solutions for challenges

As mentioned in the detailed review article [29], we can say that the main challenges are still dynamic backgrounds, registration errors and small targets. Using extra features like LBP for better performance also increases the computational cost, it is not suitable for real-time requirements of high dimensional videos. Therefore, an alternative solution might be to create a background model by only using color features and process the texture features only for the extracted candidate target regions. This allows to avoid extracting texture features for each pixel. In addition to texture features, classical methods and/or Deep Neural Networks (DNN) can be used to find a similarity score between background image and current frame for candidate target regions. Structural Similarity (SSIM) score [30] can be used to measure the similarity between image patches. As an alternative, any pre-trained CNN model could be used for feature extraction. But using a lightweight sub-network is important since it will be applied to many candidate regions. **Figure 6** shows sample detected bounding boxes with BSDOF method on PESMOD dataset. **Table 3** represents average SSIM scores between current frame and background image patches for ground truth and false positives (FP).

**Figure 6.**
*Moving object detection output of BSDOF for Pexels-Shuraev-trekking sequence.*

| Sequence name | SSIM (GT) | SSIM (FP) |
|---|---|---|
| *Pexels-Elliot-road* | 0.2569 | 0.3930 |
| *Pexels-Miksanskiy* | 0.3525 | 0.7599 |
| *Pexels-Shuraev-trekking* | 0.3511 | 0.6493 |
| *Pexels-Welton* | 0.4164 | 0.4671 |
| *Pexels-Marian* | 0.3797 | 0.3934 |
| *Pexels-Grisha-snow* | 0.4164 | 0.3875 |
| *Pexels-zaborski* | 0.4290 | 0.3691 |
| *Pexels-Wolfgang* | 0.3410 | 0.6077 |

**Table 3.**
*SSIM scores for ground truth (GT) and false positives (PF) of BSDOF method.*

Experiments with similarity comparison results show that it can be useful to eliminate some false detections caused by registration errors and illumination changes. Similarity score is expected high for false detection (no moving objects) and low for moving object regions. But, as a result of our observations, it has been observed that the similarity measure can be low in very small areas such as 5x5 pixels and in regions with no moving object. The background model can be blurred for some pixels due to registration error and/or moving background. It results low similarity score for these cases. In general, extreme wrong detections could be eliminated with a high threshold value not to lose the true detections.

Image registration errors cause possible false detection, especially for objects with sharp edges. Even if similarity comparison can help to eliminate false detection, simple tracking approaches could also be used for this issue. Historical points of each detection are stored in a *tracker list*, and detection for each frame is compared to the *tracker list*. So, tracked regions can be classified by hit count (number of detection consecutive frames) and total pixel displacements. However, it should be noted that coordinate values in the *tracker list* must be adjusted for each frame to eliminate global motion. This approach will work well if the moving target region can be extracted successfully in consecutive frames and the bounding boxes overlap with high intersection of union (IOU) value for good matching. As an alternative approach, a robust tracking method can be used but probably requires more computational cost. Targets detected with the moving object detection algorithm can be tracked with a robust tracker to obtain more precise results, and thus the tracking process continues in case the target stops.

As another approach, classical background modeling and deep learning-based methods can be used in collaboration with different processes. Our experiments show that classical methods suffer more from image registration errors, especially for fast camera movements. Therefore, the classical method and deep learning results can be combined using different strategies according to camera movement speed. Alternatively, dense optical flow with deep learning could be applied only for small patches detected by classical background modeling. In order to implement such an approach a software infrastructure in which background modeling and deep learning methods working in different processes communicate with each other and share data is essential in terms of speed. It allows us to run the processes in a pipeline logic to speed up the algorithm as shown in **Figure** 7. In the proposed architecture, process-1 applies classical background modeling approach and informs process-2 to start via zeroMQ.

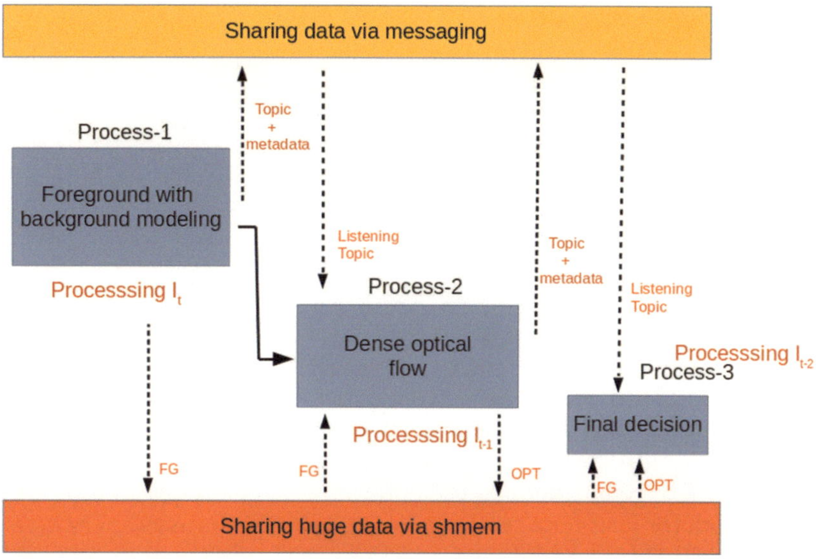


**Figure 7.**
*Software architecture to run processes in pipeline logic.*

ZeroMQ messaging library is used to transfer meta-data and inform the other processes to begin to process the frame that is ready. The foreground mask cannot be shared via messaging protocols in real-time, so that shared memory (shmem) is used to transfer this huge data between processes. Accordingly, the foreground mask is transferred to process-2 with shared memory and process-2 applies deep learning based dense optical flow only for patches extracted from input foreground mask. Finally, process-3 estimates moving target bounding boxes by processing dense optical flow output. Process-1 processes $I_t$ while process-2 processes $I_{t-1}$ with such a parallel working structure created in the pipeline logic.

## 3. Object tracking with UAV images

Object tracking is the re-detection of a target in consecutive frames after the tracker is initialized with the first bounding box as an input. It is a challenging problem for situations such as fast camera movement, occlusion, background movements, cluttering, illumination and scale changes. Tracking methods can be evaluated in different categories such as detection-based tracking, detection-free tracking, 3D object tracking, short-term tracking and long-term tracking. Detection-based tracking requires an object detector and tracking indicates assigning ID for each object. Detection-free tracking can be preferred for UAV images to handle any kind of targets and small-sized objects which is hard to detect with an object detector. As a simple approach, we can consider that we can eliminate the wrong detections after following each candidate moving object region and confirming the movement of the object with the tracker. Then we can decide for moving object with the output of the tracker. Thus, target tracking can be used in cooperation with motion detection to increase accuracy and provide better tracking.

The software architecture suggested in the previous section also seems reasonable to implement the tracking method applied after the motion detector. In this section,

| | Precision | Recall | F1 |
|---|---|---|---|
| KCF [33] | 0.4456 | 0.1214 | 0.1908 |
| CSRT [34] | 0.5006 | 0.5573 | 0.5275 |
| ECO [35] | 0.4965 | 0.5241 | 0.5099 |
| TLD [32] | 0.2460 | 0.4523 | 0.3186 |
| Re3(S) [36] | 0.4680 | 0.8030 | 0.5913 |

**Table 4.**
*Performance comparison of tracker methods on UAV123 long-term tracking sequences.*

we compare the performances of some tracker methods on UAV123 dataset [31].The dataset consists of a total of 123 video sequences obtained from low-altitude UAVs. The 20 subset images in the dataset are evaluated separately for long-term object tracking, in which targets sometimes occludes, appear and disappear, providing a better benchmark for long-term tracking. We compare performances of classical methods such as TLD [32], KCF [33], CSRT [34], ECO [35] and deep learning based method Re3 [36]. In classical methods, only TLD can handle disappeared targets in long-term tracking. Even if ECO and CSRT trackers are successful for tracking non-occluded objects, they do not have a mechanism to re-detect the object after failed. TLD can recover from full occlusion but produces frequent false positives. KCF is faster than TLD, CSRT and ECO but has lower performance. ECO and CSRT has reasonable performances except oclusion and recovering case specially important in long-term tracking. On the other hand, lightweight Re3 model can track objects at higher FPS (about 100–150 according to the GPU specifications). It allows us to track multiple objects in real-time. Average tracker performances are represented in **Table 4** for UAV123 long-term subset sequences.

Re3(S) indicates the small (lightweight) re3 model in the **Table 4** and average score shows that Re3 has the best recall score by far. In the performance comparison, the moving target detection is considered true (TP) if the intersection of union (IOU) between predicted and ground truth bounding box is greater than 0.5. Experiments show us that a moving object algorithm with tracking method support will provide significant advantages both in eliminating wrong detection and in continuous tracking.

## 4. Training CNN for moving object detection

Deep learning based solutions are an important alternative to eliminate the disadvantage of classical methods for moving object detection problem, because background modeling based methods suffer from high number of false detections. We have mentioned the deep learning based optical flow studies at the beginning of the chapter. This section summarizes the situation for supervised deep learning methods performed in the problem of moving object detection.

Deep learning based methods outperform the classical image processing based methods in CDNET dataset, but CDNET does not contain free motion images/videos. CDNET ground truths are pixel-wise masks of moving objects. FgSegNetV2 [37] is a encoder-decoder type deep neural network, and performs well on the CDNET dataset. MotionRec [38] is a single-stage deep learning framework proposed for moving object detection problem. It firstly estimates the background representation from past history frames with a temporal depth reduction block. The temporal and spatial features are used to generate multi-level feature pyramids with a backbone model. Finally, multi-level feature pyramid is used in the regressing and classification layers. MotionRec runs in the range of 2 to 5 fps depending on the selected temporal history depth from 10 to 30, over Nvidia Titan Xp GPU. JanusNet [39] is another deep network trained for moving object detection problem from UAV images. It tries to extract and combine dense optical flow and generates a coarse foreground attention map. Experiments show that it efficiently detects small moving targets. JanusNet is trained with a simulated dataset, which is generated using Unreal Engine 4. It runs at 25fps on Nvidia GTX1070 GPU and 3.1 fps on Nvidia Jetson Nano for 640 × 640 resolution images. JanusNet has also a performance comparison with the FgSegNetV2, and it shows that FgSegNetV2 cannot perform well for UAV videos due to requiring to be trained on a specific scene to work well on that scene. Considering the deep learning studies in the literature and the datasets used for training the model, it can be said that there is still a long way to go for a general-purpose supervised moving object detection method. On the other hand, classical methods can achieve reasonable results with the additional post-processing techniques and most importantly, they can work in real-time even at Nvidia modules at the edge.

## 5. Conclusions

This chapter discusses the moving object detection problem for UAV videos. We represent datasets, the performance of some methods in the literature, the challenges, and prospective solutions. For motion detection, especially background modeling-based methods are emphasized, and some post-processing methods are proposed to improve the performance as a solution to the challenges. We propose dense optical flow and simple tracking as a post-processing step with specific software architecture. Moreover, we evaluate selected trackers on a long-term object tracking dataset to analyze the performances of the trackers. Finally, we introduce some deep learning architectures and compare traditional methods in terms of general-purpose and real-life use.

## Author details

İbrahim Delibaşoğlu
Faculty of Computer and Information Sciences, Department of Software Engineering, Sakarya University, Sakarya, Turkey

*Address all correspondence to: ibrahimdelibasoglu@sakarya.edu.tr

## References

[1] Chapel M, Bouwmans T. Moving objects detection with a moving camera: A comprehensive review. Computer Science Review. 2020;**38**:100310

[2] Collins R, Lipton A, Kanade T, Fujiyoshi H, Duggins D, Tsin Y, et al. A system for video surveillance and monitoring. VSAM Final Report. 2000; **2000**:1

[3] Bouwmans T, Hofer-lin B, Porikli F, Vacavant A. Traditional approaches in background modeling for video surveillance. Handbook Background Modeling And Foreground Detection For Video Surveillance. Taylor & Francis Group; 2014

[4] Allebosch G, Deboeverie F, Veelaert P, Philips W. EFIC: Edge based foreground background segmentation and interior classification for dynamic camera viewpoints. International Conference On Advanced Concepts For Intelligent Vision Systems. 2015. pp. 130-141

[5] Zivkovic Z, Van Der Heijden F. Efficient adaptive density estimation per image pixel for the task of background subtraction. Pattern Recognition Letters. 2006;**27**:773-780

[6] Moo Yi K, Yun K, Wan Kim S, Jin Chang H, Young Choi J. Detection of moving objects with non-stationary cameras in 5.8 ms: Bringing motion detection to your mobile device. Proceedings of the IEEE Conference on Computer Vision and Pattern Recognition Workshops. 2013. pp. 27-34

[7] Zivkovic Z. Improved adaptive Gaussian mixture model for background subtraction. Proceedings of the 17th International Conference on Pattern Recognition. 2004. pp. 28-31

[8] De Gregorio M, Giordano M. WiSARDrp for Change Detection in Video Sequences. ESANN; 2017

[9] Stauffer C, Grimson W. Adaptive background mixture models for real-time tracking. IEEE Computer Society Conference on Computer Vision and Pattern Recognition (Cat. No PR00149). 1999. pp. 246-252

[10] Kim S, Yun K, Yi K, Kim S, Choi J. Detection of moving objects with a moving camera using non-panoramic background model. Machine Vision and Applications. 2013;**24**:1015-1028

[11] Zhong Z, Zhang B, Lu G, Zhao Y, Xu Y. An adaptive background modeling method for foreground segmentation. IEEE Transactions on Intelligent Transportation Systems. 2016;**18**: 1109-1121

[12] Zhong Z, Wen J, Zhang B, Xu Y. A general moving detection method using dual-target nonparametric background model. Knowledge-Based Systems. 2019; **164**:85-95

[13] Yun K, Lim J, Choi J. Scene conditional background update for moving object detection in a moving camera. Pattern Recognition Letters. 2017;**88**:57-63

[14] Yu Y, Kurnianggoro L, Jo K. Moving object detection for a moving camera based on global motion compensation and adaptive background model. International Journal of Control, Automation and Systems. 2019;**17**: 1866-1874

[15] Delibasoglu I. Real-time motion detection with candidate masks and region growing for moving cameras.

Journal of Electronic Imaging. 2021;**30**: 063027

[16] Tomasi C, Kanade T. Detection and tracking of point. International Journal of Computer Vision. 1991;**9**:137-154

[17] Fischler M, Bolles R. Random sample consensus: A paradigm for model fitting with applications to image analysis and automated cartography. Communications of the ACM. 1981;**24**: 381-395

[18] Heikkilä M, Pietikäinen M, Heikkilä J. A texture-based method for detecting moving objects. BMVC. 2004; **401**:1-10

[19] Huerta I, Rowe D, Viñas M, Mozerov M, Gonzàlez J. Background Subtraction Fusing Colour, Intensity and Edge Cues. Proceedings of the Conference on AMDO. 2007. pp. 279-288

[20] Zhao P, Zhao Y, Cai A. Hierarchical codebook background model using haar-like features. IEEE International Conference on Network Infrastructure and Digital Content. 2012. pp. 438-442

[21] Bilodeau G, Jodoin J, Saunier N. Change detection in feature space using local binary similarity patterns. International Conference on Computer and Robot Vision. 2013. pp. 106-112

[22] Wang T, Liang J, Wang X, Wang S. Background modeling using local binary patterns of motion vector. Visual Communications and Image Processing. 2012. pp. 1-5

[23] Ilg E, Mayer N, Saikia T, Keuper M, Dosovitskiy A, Brox T. Flownet 2.0: Evolution of optical flow estimation with deep networks. Proceedings of the IEEE Conference on Computer Vision and Pattern Recognition. 2017. pp. 2462-2470

[24] Huang J, Zou W, Zhu J, Zhu Z. Optical flow based real-time moving object detection in unconstrained scenes 2018

[25] Butler D, Wulff J, Stanley G, Black M. A naturalistic open source movie for optical flow evaluation. European Conference on Computer Vision (ECCV). 2012. pp. 611-625

[26] Delibasoglu I. UAV images dataset for moving object detection from moving cameras. 2021

[27] Wang Y, Jodoin P, Porikli F, Konrad J, Benezeth Y, Ishwar P. CDnet 2014: An expanded change detection benchmark dataset. Proceedings of the IEEE Conference on Computer Vision and Pattern Recognition Workshops. 2014. pp. 387-394

[28] Collins R, Zhou X, Teh S. An open source tracking testbed and evaluation web site. IEEE International Workshop on Performance Evaluation of Tracking and Surveillance. 2005. p. 35

[29] Garcia-Garcia B, Bouwmans T, Silva A. Background subtraction in real applications: Challenges, current models and future directions. Computer Science Review. 2020;**35**:100204

[30] Wang Z, Bovik A, Sheikh H, Simoncelli E. Image quality assessment: From error visibility to structural similarity. IEEE Transactions on Image Processing. 2004;**13**:600-612

[31] Mueller M, Smith N, Ghanem B. A benchmark and simulator for uav tracking. European Conference on Computer Vision. 2016;**2016**:445-461

[32] Kalal Z, Mikolajczyk K, Matas J. Tracking-learning-detection. IEEE Transactions on Pattern Analysis and Machine Intelligence. 2011;**34**:1409-1422

[33] Henriques J, Caseiro R, Martins P, Batista J. High-speed tracking with kernelized correlation filters. IEEE Transactions on Pattern Analysis and Machine Intelligence. 2014;**37**:583-596

[34] Luke A, Voji T, Zajc L, Matas J, Kristan M. Discriminative correlation filter tracker with channel and spatial reliability. International Journal of Computer Vision. 2018;**126**(7):671-688

[35] Danelljan M, Bhat G, Shahbaz Khan F, Felsberg M. Eco: Efficient convolution operators for tracking. Proceedings of the IEEE Conference on Computer Vision and Pattern Recognition. 2017. pp. 6638-6646

[36] Farhadi D, Fox D. Re 3: Real-time recurrent regression networks for visual tracking of generic objects. IEEE Robotics and Automotive Letters. 2018; **3**:788-795

[37] Lim L, Keles H. Learning multi-scale features for foreground segmentation. Pattern Analysis and Applications. 2020; **23**:1369-1380

[38] Mandal M, Kumar L, Saran M. MotionRec: A unified deep framework for moving object recognition. Proceedings of the IEEE/CVF Winter Conference on Applications of Computer Vision. 2020. pp. 2734-2743

[39] Zhao Y, Shafique K, Rasheed Z, Li M. JanusNet: Detection of moving objects from UAV platforms. Proceedings of the IEEE/CVF International Conference on Computer Vision. 2021. pp. 3899-3908

Section 2

# Methodologies

Chapter 6

# Improving Face Recognition Using Artistic Interpretations of Prominent Features: Leveraging Caricatures in Modern Surveillance Systems

*Sara R. Davis and Emily M. Hand*

## Abstract

Advances in computer vision have been primarily motivated by a better understanding of how humans perceive and codify faces. Broadly speaking, progress made in the fields of face recognition and identification has been strongly influenced by the biological mechanisms identified by research in the field of cognitive psychology. Research in cognitive psychology has long acknowledged that human face recognition and identification rely heavily on prominent features and that caricatures are capable of modeling prominent features in a multitude of ways. The field of computer science has done little to no research in the area of application of prominent features to recognition systems. This chapter discusses existing caricature research in cognitive psychology and computer vision, current issues with the practical application of caricatures to face recognition in computer vision, and how caricatures can be used to improve existing surveillance systems.

**Keywords:** face recognition, caricatures, datasets

## 1. Introduction

The word "caricature" comes from Italian for "to exaggerate" [1, 2]. As such, caricatures are artistic renderings of a human face that exaggerate prominent features while still maintaining their resemblance to the original, veridical face. Veridical is defined as the ground truth face [3]. An example can be seen in **Figure 1**. Since the 1590s, caricatures have been considered a humorous art form, meant to either entertain or humiliate, depending on context. In the United States, caricatures rose in popularity following the American Civil War, giving rise to our modern-day interpretation of the art form. At first, these images were used to mock political leaders in an effort to humorously instill political ideology [2].

Today, many people consider caricatures to be "fun" drawings. However, the fields of psychology and neuroscience have recognized the potential application of

**Figure 1.**
*A veridical image (photo) and caricature of David Tennant. (a) veridical image of David Tennant, (b) caricature image of David Tennant.*

caricatures for improving automated face verification and identification systems in recent years. Not only can caricatures be identified more often and faster than veridical images by humans [4], but they can improve the accuracy of low-resolution face verification in elderly populations [3]. Studies have also shown that introducing faces using caricatures, rather than veridical images, results in higher recognition and verification overall [5–8]. These works also show that facial exaggeration in a caricature past a certain point can actually decrease recognition performance over veridical images [9]. Each of these factors makes caricatures the ideal model for exploring how humans perceive and codify faces under nonideal conditions. In this chapter, we discuss how machine learning can leverage this biologically inspired recognition mechanism to improve surveillance systems. We discuss data collection methods and possible system architectures and show that caricatures can be used to train more robust face recognition systems.

## 2. Caricatures in cognitive psychology

Past advances in automated face recognition and verification have been driven by face perception research advancement in the field of cognitive psychology [1]. Human facial recognition is not negatively impacted by variation in pose, lighting, or resolution, unlike automatic systems [10–12]. Additionally, research has shown that human facial recognition of familiar faces is consistently better than automated systems [4]. Thus, we propose using caricature research from the field of cognitive psychology to construct surveillance systems that are robust to changes in angle, lighting, and accidental exaggeration.

The study of facial recognition in cognitive psychology has two different schools of thought: holistic and nonholistic. Each has research to support it, though we argue that the holistic approach has the greatest probability of being applied to automated facial recognition systems, and this is supported by past research [14]. Holistic face recognition research contends that faces are stored in human memory using the relationship between all features in a face, while nonholistic research argues that a single prominent facial attribute is enough to perform face recognition [15]. The difference between the two can be thought of as holistic valuing the sum of the parts of the face to create an overall model, while nonholistic values single prominent features. Because holistic face recognition relies on facial feature relationships, the relationships can be divided into two categories: featural and configural. Featural facial feature attributes look at the general structure of the facial feature, while configural focuses on the relative placement and distance between features. An example of the difference can be seen in **Figure 2**, and is taken from ref. [13].

A literature survey [16] found that most facial recognition performed by humans appears to use a holistic approach, considering each attribute in relation to the other available facial attributes. Other work has looked at face recognition in holistic and nonholistic settings to compare the possible way humans represent faces in memory [15]. They found that participants could more accurately identify unique facial attributes when the attribute was presented with the whole face as context, rather than the isolated facial attribute. For example, when participants viewed a nose on its own, they were less likely to identify that facial feature as prominent. However, if the participant viewed that same nose on the face that it belonged to, they had an easier

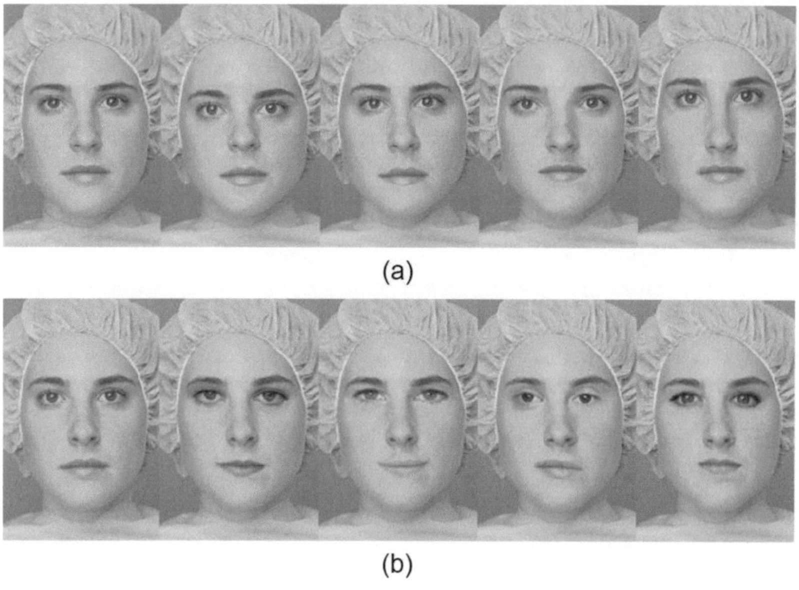


**Figure 2.**
*An illustration of the difference between configural (top) and featural (bottom) feature relationships taken from ref. [13].*

time identifying the nose as prominent. This supports the theory that humans represent faces holistically, and that they need to compare facial features to each other in order to determine which features are most prominent for that identity. Additionally, they found that if faces were inverted, recognition accuracy drastically decreased whether the part was presented with the face as context or on its own. This implies that representations in memory of human faces are strongly correlated with orientation and configural properties of facial attributes.

In order to test the importance of configural properties, another study [14] manipulated the distance between eyes, and participants were tasked with performing face recognition using distance-altered eyes. Each participant was presented with the altered eyes in (1) the original face, (2) a slight alteration of the original face, and (3) isolation. The authors found that face recognition was best using the original face, followed by the slightly altered original face, and the worst performance was when the distance-altered eyes were presented in isolation. Related to prominent feature recognition, the authors also found the configuration of the original face with the altered-eye distance resulted in lower accuracy rates in recognition of the nose and mouth features, even though the nose and mouth features had not been altered. This further supports the holistic school of thought, that is, humans learn faces holistically, and understanding the face depends both on featural and configural information.

Research has shown that participants are able to more quickly identify faces using simple line drawings of caricatured faces as compared to veridical faces [7]. This same study also suggested that caricatures can be used to better understand how humans represent faces using prominent facial features. Specifically, they found that faces and caricatures are stored in memory using the deviation of a prominent feature from the normal presentation of that feature. The authors call this norm-based coding. Another study found that the improvement in face verification rates is not specific to caricatures but is most likely caused by memory retention of facial features that deviate from the average [3]. This implies that human face encoding is closely affiliated with prominent facial features. In another study, Rhodes performed a series of experiments that tested the possible relationship between configural-based coding and norm-based coding using caricatures [8]. They found that the configural-based coding that is necessary for veridical face recognition is not necessary for caricature recognition. This implies that (1) caricature/veridical face pair recognition relies on a memory mechanism that is independent of the face/facial feature pair recognition, (2) caricature/veridical face pair recognition relies on norm-based coding, and (3) the approach to performing veridical/veridical, caricature/caricature, and caricature/veridical image recognition should be different due to the difference in coding methods.

Research has found that caricatures are accurately identified more quickly and more often than veridical images, with caricatures of familiar faces being recognized with the best accuracy [6]. Additional studies have found that caricatures of unfamiliar faces also improved verification rates by approximately 30%. Furthermore, above a certain rate of exaggeration, caricature verification is actually hindered; in other words, caricatures need to have a reasonable resemblance to the original face [9]. Past work also found that using caricatures led to better recognition of unfamiliar faces across the entire human lifespan, that it improved low-resolution face verification in older adults, and that face verification of other races also improved [3]. Each of these studies indicates that there is a link between human facial recognition, prominent features, and the general configuration of facial features.

Cognitive psychology defines facial features as either internal, such as the eyes, nose, or mouth, or external, such as hair or chin [17]. An example can be seen in

**Figure 3**. Past work has shown that familiar faces are more accurately identified if internal features were used, rather than external [17]. Feature type did not have an effect on identifying unfamiliar faces. The authors argued that the manner in which faces are modeled and stored in memory is different for familiar and unfamiliar faces, and thus, their treatment in facial recognition should be different. However, another study [18] found that internal and external features both activate similar face-selective regions of the brain, though internal features result in a greater response. Both of these works [17, 18] found that internal features were more important for familiar faces. Additionally, the study found that altering just the external features resulted in a decrease in identification accuracy, regardless of whether the face was familiar or unfamiliar. This indicates that the internal and external features interact with each other to create a holistic representation in memory and that internal and external features are likely of similar importance in machine learning applications to understanding faces. This is of particular importance to the application of caricatures to surveillance system construction because most face datasets are constructed of famous individuals; however, fame is not consistent across countries and cultures. For example, Fan Bingbing is a famous actress in China, but she is not nearly as well known in the United States. Since face recognition datasets are often comprised of a variety of celebrities from around the world, approaches to automated face recognition should not assume familiarity with the subjects in the dataset in order to make the approach relatable to human face recognition processes. Put simply, since the normal participant in a cognitive psychology study is unlikely to be familiar with every individual in a face recognition dataset, any automated system built for facial recognition should not assume familiarity with the subject.

Experiments surrounding how faces are learned over time have also been conducted. Research has shown that after a single view of a face, recognition from a different viewpoint was better using internal features rather than external features [19]. Additionally, the study found that after repeated exposure to a face, removing external features that change with high frequency, such as hair, resulted in better identification when a face was viewed from a different viewpoint. These results suggest that providing too much inconstant information to an automated face recognition system can result in a reduction of recognition capability. This means that the

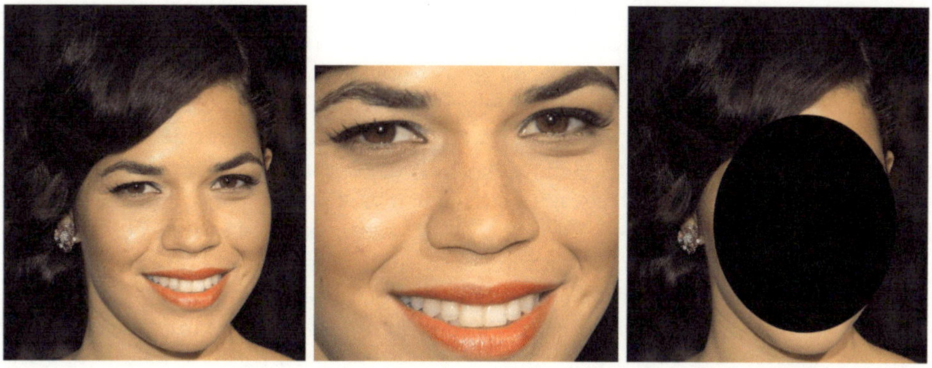

(a) (b) (c)

**Figure 3.**
*Examples of the difference between internal and external features. The original veridical image is shown in (a), internal features in (b), and external features in (c). (a) cropped full face, (b) internal facial features, (c) external facial features.*

method in which images of the face are cropped, image background, lighting, and even changes in hairstyle and makeup all likely have a significant effect on automated recognition. Other work [20] found that repeated exposure to the same face within different contexts—variations in pose and lighting—resulted in better facial recognition than if subjects simply viewed the same image of the face over and over again. This indicates that exposure to unique images has an effect on how faces are learned and retained in memory, which means that in order for an accurate facial representation to be built, automated systems need to utilize a variety of images for each identity and repeated exposure.

## 3. Caricatures in computer science

Today's automated surveillance systems rely on identity matching in some latent space [21]. We argue that the use of caricatures would better allow these systems to describe faces and prominent features, thus allowing for greater variability in pose, lighting, etc. Research shows that current automated face recognition and verification systems perform better than human recognition and verification. However, that same research suggests that automated systems only perform better on carefully curated datasets [4, 21, 22]. In other words, automated systems cannot handle images that are not taken under ideal lighting, pose, and resolution conditions. Humans, on the other hand, are capable of recognizing faces under nonideal conditions. Many surveillance systems require face alignment in order to achieve state-of-the-art results [23, 24]. This means that they work best on frontal facing images [25–27], but humans do not need a frontal facing image to perform recognition. In fact, many caricatures exaggerate face angles, and humans are still able to perform recognition with them. To improve existing automated systems, we discuss using caricatures to construct surveillance systems that are robust to changes in angle, lighting, and accidental exaggeration, and the existing research in computer science that has already leveraged these images.

Past work in automated face recognition in computer science belongs to one of two system types: traditional machine learning or deep learning. Traditional machine learning techniques to perform face recognition include using deep belief networks [28], metric learning [29, 30], and dimensionality reduction via principal component analysis (PCA) and/or linear discriminant analysis (LDA) [31]. With the rise of better hardware and GPU cycling, deep learning has become the standard approach. Typical approaches use convolutional neural networks (CNNs) [32] or autoencoders [33] and may be combined with more traditional methods to increase performance [34]. Most methods try to increase interclass margins while decreasing intraclass margins, so that distinct class clusters are created in high-dimensional feature space [35]. The recognition task is complicated by pose variance, lighting changes, and changes in an individual's appearance [36, 37], as mentioned before. Nguyen *et al.* [38] proposed a representation learning method to overcome the issues caused by recognition under nonideal conditions. They found that the cosine similarity between images can be used to improve face recognition under nonideal conditions. Past research has also shown that soft biometrics, such as the use of prominent facial features or hairstyle, can be used to improve facial recognition technology [31, 39].

Research in the area of feature learning and architecture found that facial recognition methods can be improved by utilizing ResNet CNN architectures, rather than VGG [40]. The same study discusses methods of face detection, facial alignment, and how to determine what ResNet structure is best for a selected dataset, with significant performance improvements on standardized datasets with wide variance. Unfortunately, the vast majority of work in the area of face recognition is dataset-dependent, and using proposed methods on other datasets results in an unexpected behavior [4, 21, 22], which we discuss in Section 4.

Though cognitive psychology has shown that the use of caricatures improves human recognition, work in computer science using caricatures for face recognition is rather limited. As deep learning representations for face recognition have become more accurate, face generation systems have been proposed, typically using a generative adversarial network (GAN) [42–44]. GANs are a class of deep generative models [10]. To backpropagate loss through the GAN, the input to the system must be differentiable [45]. While using a GAN can be quite successful, it can also lead to mode collapse [46] and vanishing gradient behavior [47]. Additionally, while the initial GAN results appear promising at first glance, the authors typically only report their best results and neglect to show that the vast majority of generated images are nonsensical [48]. An example of an image that is not representative of its target identity is shown in **Figure 4**. One approach to enforcing differentiability, so that better images are generated, is to use a kernel-based moment-matching scheme over a reproducing kernel Hilbert space (RKHS) [49]. This forces the real and generated images to have matched moments in the latent-feature space, and helps combat mode collapse while encouraging images that are descriptive and varied [49].

Despite these limitations, recent work generates a caricature from veridical images using GANs [41, 50–52], but does not try to understand or utilize caricatures to improve verification or recognition. Some work attempts to exploit caricatures to improve verification. However, the dataset is small and does not use modern deep learning methods [53]. Work in verification and recognition improvement using caricatures, rather than caricature generation, is relatively new and not well explored. Past work in the field [54] introduced a method to extract facial attribute features from photos but required manual labeling of facial attribute features on caricatures, which is time-consuming. Furthermore, the study computed feature importance

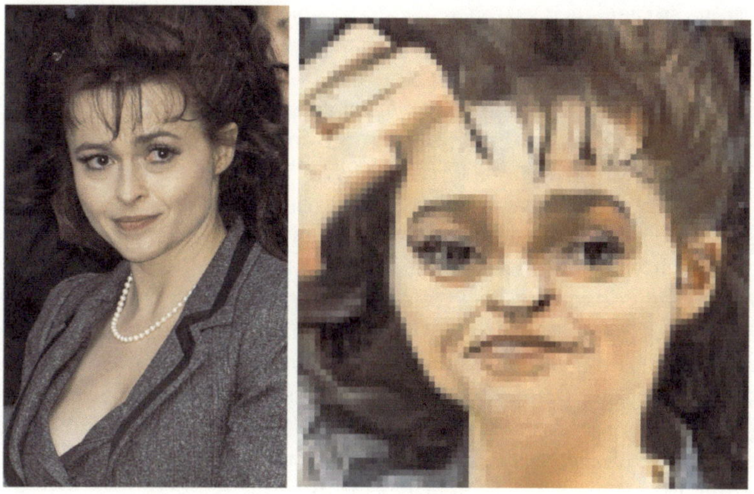

**Figure 4.**
*An example of a poorly GAN-generated caricature produced by WarpGan [41], one of the current state-of-the-art caricature generation systems. Note that this caricature (right) is not identifiable as Helena Bonham Carter (left).*

using genetic algorithms, which are extremely slow compared to deep learning. In the field of cognitive psychology, [55] showed that facial recognition improves when PCA is applied to all of an identity's images and then averaged. This indicates that the (1) human memory holds the average of a person's face after multiple exposures; and (2) PCA is one method that might be applied when creating an automated face recognition system. The most comprehensive published work in automated caricature verification is WebCaricature [56], which provides an end-to-end framework for face verification and identification using caricatures, though we discuss in Section 4 the use of flawed data in their study.

Though caricatures have not been widely used in automated face recognition systems, facial attribute recognition is a well-researched task in the field [57–62]. Recent research has focused on performing attribute recognition, and introducing new datasets and deep learning frameworks [63–66]. The current state-of-the-art facial attribute prediction methods include "Walk and Learn," which pretrains a network on face verification data and then fine-tunes it on attribute recognition [65], as opposed to pretraining on object data [67]. Work has also shown that dataset imbalance, which we discuss in Section 4, can be ameliorated by using a multi-task network with the mixed objective loss [66]. Attribute relationships have also been used within deep neural networks to improve prediction [64]. Unfortunately, current work is focused on facial attribute identification and prediction. To date, there has not been any work in using *prominent* facial features to perform recognition, despite the fact that research in cognitive psychology has shown that human recognition relies on prominent facial features (Section 2). Thus, we argue that existing surveillance systems could be improved by creating systems capable of using prominent facial features so that models are better trained to focus on the same features that humans use to identify faces.

## 4. Data collection methods

From the perspective of this chapter, we care about the application of caricature data to improve surveillance systems. Generally speaking, surveillance systems have at least one frame with an un-exaggerated snapshot of identity, similar to a photo. Therefore, caricature research typically constructs datasets by collating a caricature set and a matching identity real photo set. Past research has focused on curating datasets with as many images as possible using web scraping [49, 50, 56]. For real images, there are existing methods to remove duplicate images and images of low quality. Unfortunately the same is not true for caricatures. Since the field is relatively new, systems have not yet been built to recognize image duplicates, and even if one was constructed, it would not handle the issue of under-exaggeration or representation fidelity. Of the previously cited works, none ensure that the image is of acceptable quality, images in the caricature group are actually caricatures and not some other form of art, and that images are actually of the target identity [49, 50, 56]. In some cases, datasets inaccurately incorporate character representations of an identity, rather than the actual identity; for example, the WebCaricature dataset [56] inaccurately labels general images of Harry Potter, a cultural icon, as Harry Potter rather than gathering images of Daniel Radcliffe. This introduces a high degree of variability, as the character "Harry Potter" is not always depicted as Daniel Radcliffe, just as Daniel Radcliffe is not always seen portraying Harry Potter (**Figure 5**). Each of these conditions is critical to creating a dataset that creates an accurate caricature

**Figure 5.**
*An instance of the character representation of a fictional character (Harry Potter) not matching the affiliated actor (Daniel Radcliffe). (a) veridical image of Daniel Radcliffe. (b) pop culture representation of Harry Potter.*

**Figure 6.**
*Examples of variation in caricature representation of the same person (Patrick Stewart) taken from [56]. The left-most image is the photo (veridical face) and subsequent images are caricatures. Note that while there is wide variation in the representation of the veridical image, the identity of each of the caricatures is still obvious.*

representation of supplied identities. In a deep learning system, data quality directly affects train and test performance [67, 68]. This means that ensuring that data is representative of each identity is exceedingly important; otherwise, the recognition task becomes unnecessarily more difficult and possibly more biased.

The caricature recognition task is additionally complicated by the fact that caricatures are artistic renderings. This means that artists may choose to exaggerate some facial features over others (**Figure 6**). Furthermore, as an artistic medium, classifying an image as a "caricature" as opposed to some other art form like painting can be difficult, as shown in **Figure** 7. These two issues highlight an important core issue in data collection for caricature trained systems: (1) caricatures should be caricatures and not some other art form, and (2) the caricature should actually resemble the target identity. Since we propose using caricatures to construct surveillance systems that are robust to changes in angle, lighting, and accidental exaggeration, caricatures should still be fairly exaggerated, and using caricatures with variation in style and degree of exaggeration will improve a surveillance system's ability to accommodate large exaggerations that typically hurt performance in the existing state-of-the-art systems. Because caricature drawing is an artistic medium, the same person can be portrayed with a wide array of variations in facial features that are over or under-exaggerated. An example of this can be seen in **Figure 6**.

Unfortunately, the construction of a good caricature dataset is slow and labor intensive. Each caricature needs to be assessed for quality, and currently, the methods

**Figure 7.**
*Images from WebCaricature [56] that are not of acceptable quality to be included in a computer vision dataset. The first row contains images where the identity is not immediately obvious without knowing who the person is. The second row contains images where the image is not a caricature, but rather a painting, drawing, or cartoon.*

to do that are manual. That means checking every caricature of resemblance to the target identity and art style. Additionally, many state-of-the-art surveillance systems are reliant on facial landmarks, which can already be inaccurate in normal photos [69, 70]; this inaccuracy is exacerbated in caricatures, particularly caricatures with a high degree of exaggeration across internal facial features.

It is also critical that datasets are constructed in a way that limits bias as much as possible, as any dataset bias will be trained into a surveillance system. For example, in 2015, Google released an image labeler that had been poorly trained, so that it mislabeled human faces as gorillas [71]. Company representatives later acknowledged that this example of racism was caused by data the system was trained on.

Since many caricatures reflect cultural norms by trying to exaggerate consistent prominent features, racist interpretations are more likely. For example, because "Bruce Lee" is an Asian-American man, many racist caricatures overly exaggerate the degree of eye closure and inferred mouth pout. An exaggeration is considered racist when it does nothing to improve the machine representation of the target identity while enforcing stereotypes that exist in popular culture (**Figure 8**). Additionally, since most surveillance systems see a variety of genders, races, and angles, it is important that the dataset used to train the surveillance system is as representative as possible. The ACLU has pointed out that existing surveillance systems are more prone to misidentifying women and people of color [72]. Additionally, many police departments use mugshots to create their databases, which perpetuates the issue of racism since people of color are up to four times as likely to be arrested for the same crime perpetrated by Caucasian suspects [72]. This means that most police surveillance systems use data sets that are overwhelmingly comprised of citizens of color, making it easier to identify them than Caucasian citizens [72]. Additional work by Buolamwini and Gebru in 2018 found that datasets curated by sources other than law enforcement are composed of overwhelmingly white male subjects. This data imbalance leads to high accuracy in identifying white male subjects, but high rates of misidentification of women and people of color, and especially women of color [73]. The US Department of Commerce later reported findings consistent with Buolamwini and Gebru [74]. In a surveillance system, particularly surveillance systems used by law enforcement,

(a) (b)

**Figure 8.**
*An instance of caricaturists exaggerating racist components of Bruce Lee's features that do nothing to enhance the identifiability of the image. (a) veridical image of Bruce Lee. (b) racist representation of Bruce Lee.*

**Figure 9.**
*In some cases, veridical image and caricature image identity may be difficult to distinguish, as shown by Katy Perry (left) and Zooey Deschanel (right) in this figure. Thus, it is imperative to introduce a significant amount of quality data to allow a surveillance system to differentiate between similar faces.*

misidentification can have life-long impacts on a suspect's quality of life and likelihood to be reincarcerated.

Previous works have focused primarily on gathering as much data as possible [49, 50, 56], and while there is certainly a data problem in the machine learning field, the issue of bad data is far greater when constructing a system meant to surveil. In order for systems to learn accurate representations, those systems must be trained on accurate representations. That is, if we want a system that recognizes that two similar-looking people, such as Katy Perry and Zooey Deschanel, are different (see **Figure 9**), we need to supply a significant amount of representative data to do that, and that applies to every race, gender, and age. However, datasets gathered using a web scraper en masse tend to use celebrities, and celebrities in western culture are typically young, Caucasian, and attractive. If data were to be gathered with complete disregard for dataset balance, the face representation constructed by that system would likely perform well at identifying young, attractive, Caucasian people, and struggle with images of anyone that does not fit that description [67], and this is supported by past research [75]. While techniques like data balancing exist [64, 76], those techniques are not typically capable of fully handling the bias present in a dataset. Thus, datasets should be constructed with as much balance between gender, age, ethnicities, and image type (caricature vs veridical) as possible to ensure that the trained system is as fair as possible, *especially in systems meant to have any applicability to law enforcement*. This can be difficult to do, depending on the dataset content and availability of applicable data on the internet.

Since caricatures are a unique data source, gathering relevant, representative data is made even more difficult. Currently, the largest publicly available dataset for caricature verification and recognition is WebCaricature [56]. We have already outlined why the mindset of "quantity over quality" is detrimental to creating a fair recognition system. The WebCaricature dataset [56] illustrates this point well. WebCaricature consists of 6,042 caricatures and 5,974 veridical images over 252 identities. At a cursory glance, this dataset seems like a great resource just due to its sheer size. However, we find that there are many quality issues with the dataset itself, examples of which can be seen in **Figure 7**. First, the dataset does not bother checking that images fairly represent the target example. In other words, the target identity of each caricature is not immediately clear. Because these caricatures are not representative, they should not be included in the dataset. Second, both the caricature portion and photo portion of the dataset contain images that are not of their respective type. For example, there are multiple caricatured images that are a drawing, cartoons, or veridical images incorrectly labeled as a caricature. Third, there are many instances where the dataset contains duplicate images or images that are not of the target identity. Fourth, the authors did not collect a dataset that was balanced in terms of gender, ethnicity, or age, making it (and any system trained on it) inherently biased. Fifth, and finally, there are many included identities that have dozens of veridical images and only a single caricature image. This introduces a bias toward photo representations into the dataset and any system that uses it. After careful analysis, it becomes clear that the WebCaricature dataset's focus on quantity has led to a marked decrease in quality that would unduly bias any surveillance system that uses it.

Thus, we propose that the following list of questions be used to construct future caricature datasets:

1. Are there a significant number of images (caricature AND veridical) of this identity to create an accurate representation of this person?

2. Are images being used as caricatures actually caricatures, and do the caricatures accurately represent the identity without incorporating racist or cultural elements?

3. Factoring in all identities, is there as much balance across identity race, gender, and age as possible?

4. Factoring in all images, is there as much balance across identity race, gender, and age as possible?

We concede that most publicly sourced datasets from Western culture will have an easier time collecting images of white individuals. That means that maintaining race balance will restrict the number of images of Caucasian subjects that can be collected since a roughly equal balance is necessary to ensure fairness. In terms of dataset size, this means sacrificing quantity for quality in the interest of creating a fair surveillance system.

## 5. Using caricatures for prominent feature recognition in surveillance systems

As discussed in Section 3, existing research does not address prominent facial feature recognition, despite the fact that cognitive psychology has been trying to better identify them for decades. The most common approach to designing the system architecture for face recognition is to first detect any faces in the image and then to landmark that image. The landmarks are then given as input to some sort of machine learning algorithms, such as a deep belief network [28], convolutional neural network [56], or genetic algorithm [53]. We believe that this same generalized process can continue to be used, so long as the field addresses the gap between prominent feature research in cognitive psychology and computer vision. This can be accomplished by using caricatures to better model prominent feature exaggeration.

Future work using caricatures should seek to address this critical gap in facial recognition by doing the following:

1. Improve prominent feature recognition and labeling using caricatures.

2. Utilize prominent feature recognition to better train multi-task surveillance systems that leverage unique facial attributes.

3. Generate high-fidelity caricatures with a strong resemblance to the veridical image.

We address each of these points below, with suggestions for courses of future research.

Future research should use landmarking on caricature and veridical images to measure feature deviation from the average. Controlling for well-known conditions that affect feature size, such as gender and ethnicity, measuring configural and size properties of each facial feature can provide insight into what an image's prominent features are. For example, Helena Bonham Carter's eyes are large when compared to other celebrities, and they are also large in comparison to the rest of her facial features. Landmarks can be used to quantitatively analyze the difference in the size of relative features and the deviation from average in order to identify prominent features. It is worth noting that research that quantitatively analyzes feature size and shape *must* control for gender and ethnicity in order to create better models; past medical research has shown that nose shape, for example, is highly correlated with race [77]. By controlling for variables, such as race and gender, systems trained to recognize prominent features can be better attuned to small differences between features in different subjects. We warn that systems that do not implement this control into their experiments are likely to miss fine-grained feature differences, and may perpetuate bias, which is an obvious downside to the use of prominent facial features if they are not used carefully. Providing landmark data to a simple machine learning model, such as a support vector machine (SVM), or to a deep convolutional neural network should provide a baseline method for prominent facial feature recognition. This baseline should be simple to implement and is a first step in improving prominent feature recognition and labeling. Preliminary results using simple models may look fairly underwhelming because it is unlikely that they will optimally handle a large amount of landmarking data coming; however, results that are better than chance will indicate that prominent feature usage is worth pursuing.

The developed prominent feature recognition method can be used to train surveillance systems that are capable of leveraging prominent facial features. We argue that by using prominent facial feature labels and this prominent feature methodology, deep learning models used in surveillance can improve their performance. Additionally, prominent feature recognition methods can be used as an additional task in existing surveillance systems, which should ultimately make a more robust, less overfit model [31, 78, 79]. This second step is critical to better mimicking human face perception and should improve most recognition and surveillance systems. Again, if prominent features are used to train a surveillance system, race and gender need to be controlled as part of the proposed multitask network so that the system is not unintentionally biased.

After a model is in place to identify prominent features, the identified features from a veridical image can be used to better train existing GAN models used for caricature generation. This will create caricatures with higher fidelity to the veridical image. These generated images, can, in turn, be used to ameliorate the data quantity problem that most surveillance systems have.

We also note that the use of caricatures to improve existing system architectures may prove difficult at first, especially if collated datasets are relatively small, there may just be a data abundance problem. Therefore, it is imperative that large, well-constructed datasets be created prior to any landmarking or architecture improvement. Additionally, initial research in caricature usage will likely prove to be slow, since all data will need to be manually landmarked until a proper landmarking model is devised for caricatures. Aside from data abundance and the time necessary to manually create the dataset and landmarking systems, it is also likely that initial systems, no matter how well controlled, may end up slightly biased and better at identifying identities of specific races, genders, and ages. This is simply due to the fact that it is easiest to find existing caricature data through web scraping, and web scraping will inherently lead to more images of celebrities, which will be culturally skewed in favor of one race over another. In order to control this, data augmentation methods appropriate to caricature utilization should be looked at. Finally, we note that the general subjectivity of caricature acceptability, as discussed in Section 4, may lead to a level of variation across curated datasets and unintentional bias.

Given that most computer vision advancements have been made by a better understanding of human perception [1], we argue that utilizing cognitive psychology's findings about caricatures to our advantage will result in more robust computer vision systems, and in turn, more robust surveillance systems. By developing a method to identify prominent features, surveillance systems can better leverage the same mechanisms that human face perception uses to improve recognition.

## 6. Conclusion

Human face recognition relies on the use of prominent facial features. We outline past research in cognitive psychology that should be leveraged to improve surveillance systems. In particular, we argue that the use of prominent facial features is critical to better modeling human face perception. In addition, Section 2 discusses the importance of using a holistic face model and internal facial features to construct robust recognition systems. In Section 3, we discuss past research in computer science that leverages the use of caricatures. We note that research in this area is hindered by the lack of study of prominent facial features, and that, in most cases, existing research in computer science that uses caricatures is rather limited and of low quality. Next, we discuss the importance of collecting a dataset that is not only large but also balanced (Section 4). We argue that dataset balance in terms of gender, race, age, and image type is critical to limiting bias within surveillance systems trained on these datasets and discuss the past research that supports our stance. Additionally, we provide a series of guidelines for caricature dataset generation, so that future caricature datasets are of acceptable quality for use in surveillance system training. Finally, we outline the ways in which caricatures can be used to improve facial recognition systems. In particular, we argue that improved prominent feature labeling and recognition is critical, so that these features can be used to better train multitask surveillance systems.

## Acknowledgements

This material is based upon work supported by the National Science Foundation under Grant IIS-1909707.

## Author details

Sara R. Davis and Emily M. Hand*
University of Nevada, Reno, Reno, NV, USA

*Address all correspondence to: emhand@unr.edu

## References

[1] Scheirer WJ, Anthony SE, Nakayama K, Cox DD. Perceptual annotation: Measuring human vision to improve computer vision. IEEE Transactions on Pattern Analysis and Machine Intelligence. 2014;**36**(8): 1679-1686

[2] Wright T. A History of Caricature and Grotesque in Literature and Art. Virtue Brothers; 1865

[3] Dawel A, Wong TY, McMorrow J, Ivanovici C, He X, Barnes N, et al. Caricaturing as a general method to improve poor face recognition: Evidence from low-resolution images, other-race faces, and older adults. Journal of Experimental Psychology Applied. 2019; **25**(2):256-279

[4] Sun YK, Wang X, Tang X. Deep learning face representation from predicting 10,000 classes. In: 2014 IEEE Conference on Computer Vision and Pattern Recognition. 2014. pp. 1891-1898

[5] Michael B. Lewis. Are caricatures special? evidence of peak shift in face recognition. European Journal of Cognitive Psychology. 1999;**11**(1): 105-117

[6] Mauro R, Kubovy M. Caricature and face recognition. Memory & Cognition. 1992;**20**(4):433-440

[7] Rhodes G, Brennan S, Carey S. Identification and ratings of caricatures: Implications for mental representations of faces. Cognitive Psychology. 1987; **19**(4):473-497

[8] Rhodes G, Tremewan T. Understanding face recognition: Caricauture effects, inversion, and the homogeneity problem. Visual Cognition. 1994;**1**(2–3):275-311

[9] Alex H, Hancock PJB, Kittler J, Langton SRH. Improving discrimination and face matching with caricature. Applied Cognitive Psychology. 2013; **27**(6):725-734

[10] Goodfellow IJ, Pouget-Abadie J, Mirza M, Xu B, Warde-Farley D, Ozair S, Courville A, Bengio Y. Generative Adversarial Networks, 2014

[11] Nguyen A, Yosinski J, Clune J. Deep Neural Networks are Easily Fooled: High Confidence Predictions for Unrecognizable Images, 2015

[12] Radford A, Metz L, Chintala S. Unsupervised Representation Learning with Deep Convolutional Generative Adversarial Networks. 2015

[13] Maurer D, Le Grand R, Mondloch C. The many faces of configural processing. Trends in Cognitive Sciences. 2002;**6**: 255-260

[14] James W, Sengco JA. Features and their configuration in face recognition. Memory & Cognition. 1997;**25**:583-592

[15] Tanaka J, Farah M. Parts and wholes in face recognition. The Quarterly journal of experimental psychology. A, Human experimental psychology. 1993; **46**:225-245

[16] Tanaka JW, Simonyi D. The “parts and wholes” of face recognition: A review of the literature. Quarterly Journal of Experimental Psychology. 2016;**69**(10):1876-1889

[17] Ellis H, Shepherd J, Davies G. Identification of familiar and unfamiliar

faces from internal and external features: Some implications for theories of face recognition. Perception. 1979;**8**:431-439

[18] Andrews T, Davies-Thompson J, Kingstone A, Young A. Internal and external features of the face are represented holistically in face-selective regions of visual cortex. The Journal of Neuroscience : The Official Journal of the Society for Neuroscience. 2010;**30**: 3544-3452

[19] Christopher A, Liu CH, Young AW. The importance of internal facial features in learning new faces. Quarterly Journal of Experimental Psychology. 2015;**68**(2):249-260

[20] Murphy J, Ipser A, Gaigg S, Cook R. Exemplar variance supports robust learning of facial identity. Journal of Experimental Psychology. Human Perception and Performance. 2015;**41**:4

[21] Novak R, Bahri Y, Abolafia DA, Pennington J, Sohl-Dickstein J. Sensitivity and Generalization in Neural Networks: An Empirical Study 2018

[22] Wang M, Deng W. Deep face recognition: A survey. Neurocomputing, 2021;**429**:215-244

[23] Zhao J, Zhou Y, Li Z, Wang W, Chang K-W. Learning gender-neutral word embeddings. CoRR, abs/ 1809.01496. 2018

[24] Abate AF, Nappi M, Riccio D, Sabatino G. 2d and 3d face recognition: A survey. Pattern Recognition Letters. 2007;**28**:1885-1906

[25] Jourabloo A, Liu X. Large-pose face alignment via cnn-based dense 3d model fitting. In: IEEE Conference on Computer Vision and Pattern Recognition. 2016

[26] Zhu X, Lei Z, Liu X, Shi H, Li SZ. Face alignment across large poses: A 3d solution. In: Proceedings of the IEEE Conference on Computer Vision and Pattern Recognition. 2016. pp. 146-155

[27] Bowyer KW, Chang K, Flynn P. A survey of approaches and challenges in 3d and multi-modal 3d+ 2d face recognition. Computer Vision and Image Understanding. 2006;**101**:1-15

[28] Huang GB, Lee H, Learned-Miller EG. Learning hierarchical representations for face verification with convolutional deep belief networks. CVPR; 2012. pp. 2518-2525

[29] Cai X, Wang C, Xiao B, Xue C, Zhou J. Deep nonlinear metric learning with independent subspace analysis for face verification. In: Proceedings of the 20th ACM International Conference on Multimedia. New York, NY, USA: Association for Computing Machinery; 2012. pp. 749-752

[30] Guillaumin M, Verbeek J, Schmid C. Is that you? metric learning approaches for face identification. In: 2009 IEEE 12th International Conference on Computer Vision. 2009. pp. 498-505

[31] Hao Zhang J. Ross Beveridge, Bruce A. Draper, and P. Jonathon Phillips. On the effectiveness of soft biometrics for increasing face verification rates. Computer Vision and Image Understanding. 2015;**137**:50-62

[32] Taylor GW, Fergus R, LeCun Y, Bregler C. Convolutional learning of spatio-temporal features. In: Daniilidis K, Maragos P, Paragios N, editors. Computer Vision – ECCV 2010. Berlin, Heidelberg: Springer; 2010. pp. 140-153

[33] Vincent P, Larochelle H, Bengio Y, Manzagol P-A. Extracting and

composing robust features with denoising autoencoders. In: Proceedings of the 25th International Conference on Machine Learning, ICML '08. New York, NY, USA: Association for Computing Machinery; 2008. pp. 1096-1103

[34] Dong Y, Lei Z, Stan ZL. Towards pose robust face recognition. In: Proceedings of the IEEE Conference on Computer Vision and Pattern Recognition (CVPR). 2013

[35] Richard O, Hart PE, Stork DG. Pattern Classification. 2nd ed. New York: Wiley; 2001

[36] Cao Z, Yin Q, Tang X, Sun J. Face recognition with learning-based descriptor. In: IEEE Computer Society Conference on Computer Vision and Pattern Recognition. IEEE; 2010. pp. 2707-2714

[37] Hu J, Lu J, Tan Y-P. Discriminative deep metric learning for face verification in the wild. In: Proceedings of the IEEE Conference on Computer Vision and Pattern Recognition. 2014. pp. 1875-1882

[38] Nguyen HV, Bai L. Cosine similarity metric learning for face verification. In: Kimmel R, Klette R, Sugimoto A, editors. Computer Vision – ACCV 2010. Berlin, Heidelberg: Springer; 2011. pp. 709-720

[39] Thom N, Hand EM. Facial Attribute Recognition: A Survey. 2020

[40] Hsiao S-H, Jang J-SR. Improving resnet-based feature extractor for face recognition via re-ranking and approximate nearest neighbor. In: 2019 16th IEEE International Conference on Advanced Video and Signal Based Surveillance (AVSS). 2019. pp. 1-8

[41] Shi Y, Deb D, Jain AK. Warpgan: Automatic caricature generation. In: Proceedings of the IEEE/CVF Conference on Computer Vision and Pattern Recognition. 2019. pp. 10762-10771

[42] Gauthier J. Conditional generative adversarial nets for convolutional face generation. In: Convolutional Neural Networks for Visual Recognition. 2014. p. 2

[43] Li M, Zuo W, Zhang D. Convolutional network for attribute-driven and identity-preserving human face generation. arXiv preprint arXiv: 1608.06434, 2016

[44] Lu Y, Tai Y-W, Tang C-K. Attribute-guided face generation using conditional cyclegan. In: Proceedings of the European Conference on Computer Vision (ECCV). 2018. pp. 282-297

[45] Wang K, Wan X. Sentigan: Generating sentimental texts via mixture adversarial networks. In: Proceedings of the Twenty-Seventh International Joint Conference on Artificial Intelligence, IJCAI-18. 2018. pp. 4446-4452

[46] Metz L, Poole B, Pfau D, Sohl-Dickstein J. Unrolled generative adversarial networks. In: 5th International Conference on Learning Representations. Toulon, France: ICLR, 2017

[47] Arjovsky M, Bottou L. Towards Principled Methods for Training Generative Adversarial Networks 2017

[48] Taphorn A. Gan and Their Chances and Risks in Face Generation and Manipulation. 2020

[49] Zhang Y, Gan X, Fan K, Chen X, Henao R, Shen D, Carin L. Adversarial Feature Matching for Text Generation. 2017

[50] Jang W, Ju G, Jung Y, Yang J, Tong X, Lee S. Stylecarigan: Caricature

generation via stylegan feature map modulation. arXiv preprint arXiv: 2107.04331 2021

[51] Chiang P-Y, Liao W-H, Li T-Y. Automatic caricature generation by analyzing facial features. In: Proceeding of 2004 Asia Conference on Computer Vision (ACCV2004). Korea; 2004

[52] Zipeng Ye, Ran Yi, Minjing Yu, Juyong Zhang, Yu-Kun Lai, and Yong-jin Liu. 3d-carigan: An end-to-end solution to 3d caricature generation from face photos. IEEE Trans Vis Comput GraphIEEE Trans Vis Comput Graph, abs/2003.06841. 2021

[53] Brendan F, Bucak SS, Jain AK, Akgul T. Towards automated caricature recognition. In: 2012 5th IAPR International Conference on Biometrics (ICB). 2012. pp. 139-146

[54] Abacı B, Akgül T. Matching caricatures to photographs. Signal Image and Video Processing. 2015;**9**:1-9

[55] Mike Burton A, Jenkins R, Hancock PJB, White D. Robust representations for face recognition: The power of averages. Cognitive Psychology. 2005;**51**:256-284

[56] Huo J, Li W, Shi Y, Yang G, Yin H. Webcaricature: A benchmark for caricature recognition. In: British Machine Vision Conference 2018, BMVC 2018, Newcastle, UK: BMVA Press; 2018. p. 223

[57] Berg T, Belhumeur PN. Poof: Part-based one-vs.-one features for fine-grained categorization, face verification, and attribute estimation. Computer Vision and Pattern Recognition. 2013: 955-962

[58] Berg T, Belhumeur PN. Poof: Part-based one-vs.-one features for fine-grained categorization, face verification, and attribute estimation. In: 2013 IEEE Conference on Computer Vision and Pattern Recognition. Portland, OR, USA: IEEE Computer Society; 2013. pp. 955-962

[59] Kumar N, Belhumeur PN, Nayar SK. Facetracer: A search engine for large collections of images with faces. In David A. Forsyth DA, Torr PHS, Zisserman A, editors, Computer Vision - ECCV 2008, 10th European Conference on Computer, Vision, Marseille, Proceedings, Part IV, volume 5305 of Lecture. Notes in Computer Science. France: Springer; 2008. pp. 340-353

[60] Kumar N, Berg AC, Belhumeur PN, Nayar SK. Attribute and simile classifiers for face verification. In IEEE 12th International Conference on Computer Vision, ICCV 2009. Kyoto, Japan: IEEEComputer Society; 2009. pp. 365-372

[61] Kumar N, Berg AC, Belhumeur PN, Nayar SK. Describable visual attributes for face verification and image search. In: PAMI. 2011

[62] Layne R, Hospedales TM, Gong S, Mary Q. Person re-identification by attributes. In Bowden R, Collomosse JP, Mikolajczyk K, editors. British Machine Vision Conference, BMVC 2012, Surrey, UK: BMVA Press; 2012. pp. 1-11

[63] Dharr S, Ordonez V, Berg TL. High level describable attributes for predicting aesthetics and interestingness. In The 24th IEEE Conference on Computer Vision and Pattern Recognition, CVPR 2011. Colorado Springs, CO, USA: IEEE Computer Society; 2011. pp. 1657-1664

[64] Hand EM, Chellappa R. Attributes for improved attributes: A multi-task network utilizing implicit and explicit

relationships for facial attribute classification. In Singh S, Markovitch S, editors. Proceedings of the Thirty-First AAAI Conference on Artificial Intelligence. San Francisco, California, USA: AAAI Press; 2017. pp. 4068-4074

[65] Liu Z, Luo P, Wang X, Tang X. Deep learning face attributes in the wild. In 2015 IEEE International Conference on Computer Vision, ICCV 2015, Santiago, Chile: IEEE Computer Society; 2015. pp. 3730-3738

[66] Rudd EM, Gunther M, Boult TE. Moon: A mixed objective optimization network for the recognition of facial attributes. In: Leibe B, Matas J, Sebe N, Welling M, editors. Computer Vision - ECCV 2016 - 14th European Conference, Amsterdam, The Netherlands, October 11-14, 2016, Proceedings, Part V, volume 9909 of Lecture Notes in Computer Science, Amsterdam, The Netherlands: Springer; 2016. pp. 19-35

[67] Cortes C, Jackel LD, Chiang W-P. Limits on learning machine accuracy imposed by data quality. In: Advances in Neural Information Processing Systems. 1994. p. 7

[68] Jain B, Patel H, Nagalapatti L, Gupta N, Mehta S, Guttula S, Mujumdar N, et al. Overview and importance of data quality for machine learning tasks. In: Proceedings of the 26th ACM SIGKDD International Conference on Knowledge Discovery & Data Mining. 2020. pp. 3561-3562

[69] Cummaudo M, Guerzoni M, Marasciuolo L, Gibelli D, Cigada A, Obertovà Z, et al. Pitfalls at the root of facial assessment on photographs: A quantitative study of accuracy in positioning facial landmarks. International Journal of Legal Medicine. 2013;**127**(3):699-706

[70] Lin J, Xiao L, Wu T. Face recognition for video surveillance with aligned facial landmarks learning. Technology and Health Care. 2018;**26**(S1):169-178

[71] Google apologises for photos app's racist blunder, July 2015

[72] Crockford K. How is Face Recognition Surveillance Technology Racist?: News & Commentary, Jun 2020

[73] Buolamwini J, Gebru T. Gender shades: Intersectional accuracy disparities in commercial gender classification. In: PMLR. 2018

[74] Patrick Gother, Mei Ngan, and Kayee Hanaoka. Face recognition vendor test (frvt) - nist

[75] Lingenfelter B, Hand EM. Improving evaluation of facial attribute prediction models. In: 2021 16th IEEE International Conference on Automatic Face and Gesture Recognition (FG 2021). Jodhpur, India: IEEE; 2021. pp. 1-7

[76] Gustavo EAPA, Prati RC, Monard MC. A study of the behavior of several methods for balancing machine learning training data. ACM SIGKDD Explorations Newsletter. 2004;**6**:20-29

[77] Suhk JH, Park JS, Nguyen AH. Nasal analysis and anatomy: Anthropometric proportional assessment in asians-aesthetic balance from forehead to chin, part i 2015

[78] Argyriou A, Evgeniou T, Pontil M. Multi-task feature learning. Advances in Neural Information Processing Systems. 2007;**2007**:41-48

[79] Ranjan R, Patel VM, Chellappa R. Hyperface: A deep multi-task learning framework for face detection, landmark localization, pose estimation, and gender recognition. CoRR. 2016;abs/1603.01249

Chapter 7

# Spatial Change Recognition Model Using Image Processing and Fuzzy Inference System to Remote Sensing

*Majid Mirbod*


## Abstract

After the advent of satellites whose job is to image the surface of the earth, a huge database of imaging data of the surface of the earth was made available to researchers in various sciences to exploit a large data set in their field of work, and the subject of remote sensing gradually came to the attention of researchers in various sciences. For example, geography, environmental science, civil engineering, etc., each analyzed the visual data of the earth's surface from the perspective of their field. According to this research, the issue of spatial change recognition and their location and calculating the percentage of changes at the ground level has been considered, and the model presented is based on machine vision, image processing, and a fuzzy interface system to reveal features. This research is in the category of applied research and finally, an application will be presented that can lead to the development of software such as Google Earth and can be added to that as an option. Another of the advantages of this model is its easy use compared to specialized software such as Arc GIS, and this is the novelty of this research.


**Keywords:** fuzzy interface system, spatial change recognition, remote sensing, image processing, remote sensing application

## 1. Introduction

This paper presents a spatial change recognition model using satellite images, image processing, and a fuzzy interface system as a remote sensing application, and it is in the applied research category. Change recognition in natural phenomena is very important for managing and preserving the environment. On the other hand, video data collection satellites have been able to produce large volumes of images from the surface of the earth and provide them to researchers, such as Google Earth. But the images taken by the satellites show the state of the earth at the time of the image capture, and to get more information about whether or not there has been a change in that area, it is necessary to compare exactly the previous images there. So, change recognition in the study area can give an idea of how to manage and control the environment in that area. Also, change recognition in remotely sensed images is an active research area [1]. Remote Sensing on the Earth's surface and change recognition mean change detection on the Earth's surface by processing images of the same geographical area acquired at different times. This field can include forest or vegetation change, forest mortality, defoliation, and damage assessment, wetland change, urban expansion, damage assessment, crop monitoring, changes in glacier mass balance, environmental change and deforestation, regeneration, and selective logging [2]. So, briefly; we refer to some past research in this regard. For example: "Land cover change detection using GIS and remote sensing techniques: A Spatio-temporal study on Tanguar Haor, Sunamganj, Bangladesh" [3], that using a classification area model. Or, "Automated unsupervised change detection technique from RGB color image", that uses coefficients correlation calculation between color signatures of each two associated pixels from two satellite images for the same area [4], The change detection model has to be insensitive to illumination brightness changes [5], and for this reason, in the proposed model, we use high-resolution grayscale images to prevent changes in light and brightness in satellite images in calculating spatial changes. Or, "Change detection in the city of Hilla between 2007 and 2015 using Remote Sensing Techniques", that using ArcGIS 10.4 software and those processes include, geometric correction, spectral enhancement, image classification, and cartographic output [6], as mentioned, one of the advantages of this model, that is its easy use compared to specialized software such as Arc GIS, and this is the novelty of this research. Or understanding patterns of vegetation change at the Angkor world heritage site by combining remote sensing results with local knowledge based on analysis stages used to extract spectral plots of pixel values in the region of interest [7], or automatic change detection of buildings in an urban environment from very high spatial resolution images using existing geo-database and prior knowledge based on the Image segmentation model that result of the ratio of the detected area was 86–90% [8], or, On a survey of change detection methods based on remote sensing images for multi-source and multi-objective scenarios was expressed, at present, spatial change recognition as a remote sensing application manifests great significance in numerous change detection applications that are shown on the chart below (**Figure 1**) [9].

Another study entitled: change detection of soil formation rate in space and time based on multi-source data and geospatial analysis techniques, estimate the dissolution rate and soil formation rate in karst areas of China and analyzed their spatial diversity has been done [10], or in another study, change detection techniques based on multispectral images for investigating land cover dynamics using image processing and mining have been investigated [11]. In another study entitled: Change detection

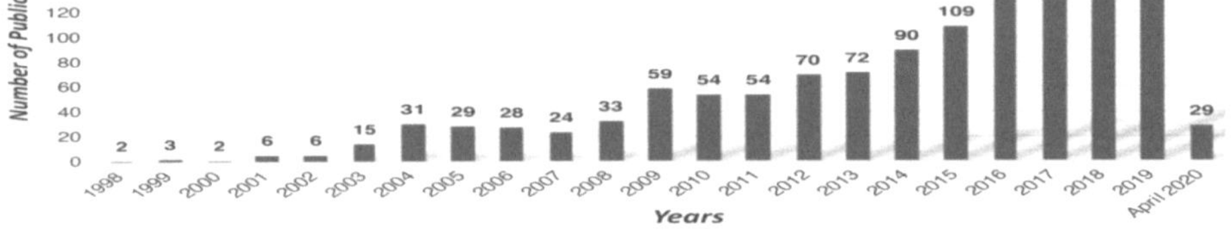


**Figure 1.**
*Published literature statistics of urban change detection according to the keywords remote sensing and urban change detection in Web of Science (total of 1283 publications) [9].*

techniques for remote sensing applications, about the distribution of change detection methods [12].

Another study using an images enhancement method, improve the accuracy of SAR[1] images to change detection [14]. Or, in another study, based on similarity measurement between heterogeneous images, change detection has been done [15]. Or based on the maximum entropy principle to obtain the final change detection map and compare with the wavelet-based textural features, plain texture difference, image difference, and log-ratio methods change detection has been done [16].

## 2. Materials and methods

### 2.1 Materials

In this research, we want to use publicly available data without the need for special software or special knowledge to collect data and test the proposed model with it. Google Earth is one of the most popular and widely available applications. And since the model is universal and the quality of the images is enhanced by using image processing techniques, it is enough to introduce two images with exactly the same spatial characteristics to the model in two different periods or to display and calculate their differences. So, for example, we take pictures of some different places in the world using the time change feature in Google Earth software and test them with the proposed model. A very important point is that the location specifications are exactly the same in capturing images, which Google Earth software has this feature. So that without changing the desired location, it is enough to change the timeline and capture two images of a place at two different times. Here are examples of images Acquisition from Google Earth from a fixed location at two different time intervals (**Figures 2–5**).

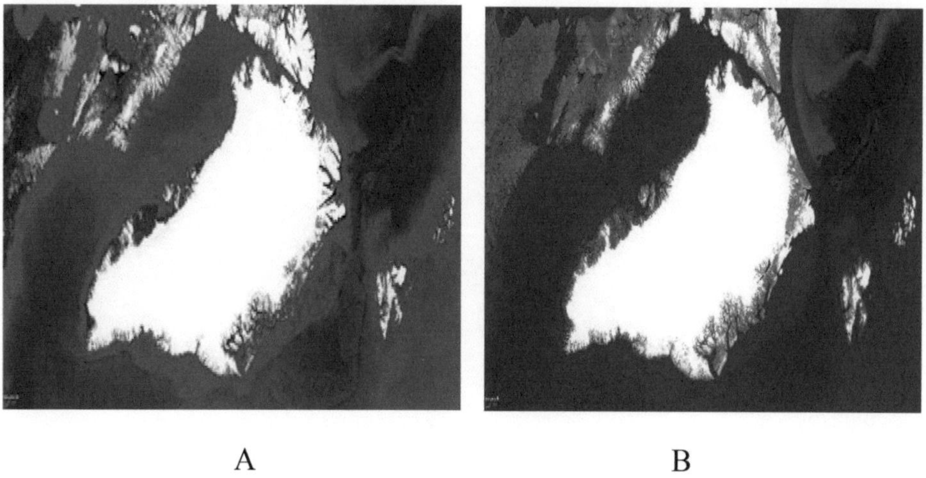


**Figure 2.**
*A: Greenland 1930, B: Greenland 2021, source and specifications of images: NOAA, US Navy, NGA, GEBCO, Landsat, 74,22′50,50 N 45,02′01,17 W, elev 2811 m, height: 3601.47 km.*

[1] Synthetic-aperture radar (SAR): synthetic-aperture radar is a form of radar that is used to create two-dimensional images or three-dimensional reconstructions of objects, such as landscapes [13].

A B

**Figure 3.**
*A: Jumeirah Palm beach 2000, B: Jumeirah Palm beach 2000, source and specifications of images: NOAA, US Navy, NGA, GEBCO, Landsat, 25,06′42,80 N 55,03′43,93 E, elev 10 m, height: 38.26 km.*

A B

**Figure 4.**
*A: Oroomiye Lake 1984, B: Oroomiye Lake 2017, source and specifications of images: NOAA, US Navy, NGA, GEBCO, Landsat, 37,08′58,85 N 45,12′31,05 E, elev 2318 m, height: 157.76 km.*

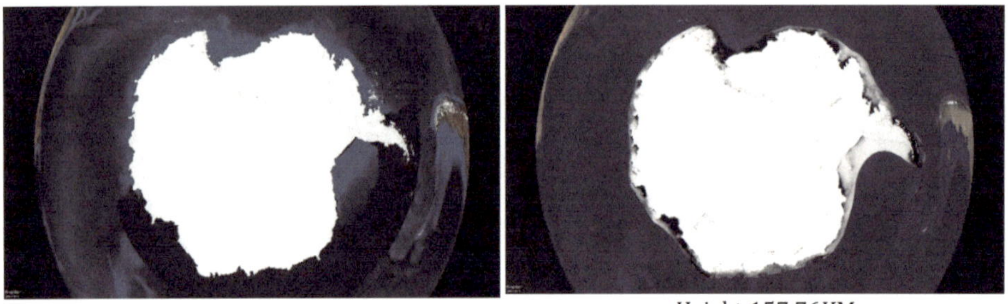

Height:157.76KM

A B

**Figure 5.**
*South Pole 1957, B: South Pole 2022, source and specifications of images: NOAA, US Navy, NGA, GEBCO, Landsat, 83,32′58,95 S 62,22′15,95 E, elev 3178 m, height: 7734.93 km.*

Similarly, any other location can be imaged and compared at two intervals, and the proposed model has no limitations, it is enough to keep the exact location specifications and the camera does not change in terms of geographical coordinates or height.

## 2.2 Methods

The basic model we use to change recognition in images has previously been used to detect changes in industrial parts, where local imaging was performed by the camera and was a type of micrography [17]. The input of the model is the images prepared in the previous section, the type of which was described. In **Figure 6**, the spatial change recognition model for remote sensing is shown.

### *2.2.1 Description of model components*

#### *2.2.1.1 Acquisition of the first and second spatial image*

The source for the acquisition of satellite spatial images with the history is the Google Earth application.

#### *2.2.1.2 Prepare spatial images data*

We used the MATLAB image processing toolbox to image mining included, convert RGB images to grayscale and converting the intensity image to double (Pre-processing and data preparation stage), and implemented other parts of the model. The reason for converting images from RGB to grayscale is reducing the data from three dimensions to two dimensions while simplifying the problem. So, in MATLAB, there is a function called "rgb2gray" is available to convert RGB images to grayscale images that we used [17].

#### *2.2.1.3 Edge detection from spatial images with different techniques*

In this part, three different methods have been used to edge detection. The reason for using these three methods is that each has its strengths and weaknesses, thus presenting different results in edge detection, which in total will lead to the model's strength in edge recognition.

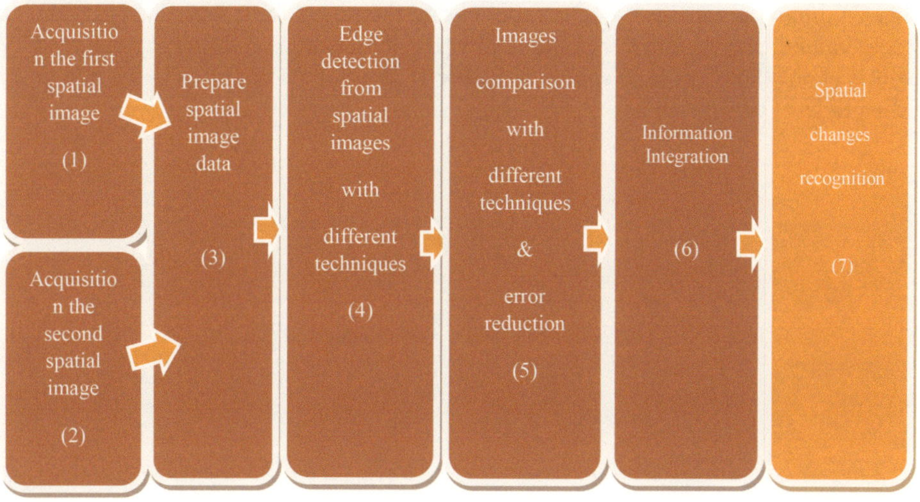


**Figure 6.**
*Spatial change recognition model to remote sensing.*

#### *2.2.1.3.1 Fuzzy interface systems to edge detection*

Briefly, the fuzzy conditions help to test the relative values of pixels that can be present in case of presence on an edge. So, the image is said to have an edge if the intensity variation between the adjacent pixels is large. The mask used for scanning the image is shown in **Figure 7** [17].

$$Gx = [-1\,1], Gy = Gx'$$

The mask is slid over an area of the spatial image and changes that pixel's value, and then shifts one pixel to the right and continues to the right until it reaches the end of a row. It then starts at the beginning of the next row and the process continues till the whole image is scanned. When this mask is made to slide over the image, the output is generated by the FIS based on the rules and the value of the pixels [17]. In summary, the steps for using a fuzzy inference system are as follows: a) Crisp spatial images for fuzzified into various FS, having conventional crisp membership functions i.e. Black and White. b) Firing strength is calculated using fuzzy t-norms operators. c) Fuzzy rules are fired for each crisp spatial image. d) Aggregate resultant output FS for all fired rules is achieved by using the max operator (s-norm). e) De-fuzzification using the Centroid method. f) The crisp output is the pixel value of the output image i.e. one containing the edges, and black and white regions. g) The first derivative is performed on the image output from FIS after the application of the noise removal algorithm. h) Further refinement is performed by the second derivative and noise removal [17].

#### *2.2.1.3.2 Sobel's operator to edge detection*

The Sobel operator sometimes called the Sobel–Feldman operator or Sobel filter is used in image processing and computer vision, particularly within edge detection algorithms where it creates an image emphasizing edges [18]. Technically, it is a discrete differentiation operator, computing an approximation of the gradient of the image intensity function. At each point in the image, the result of the Sobel-Feldman operator is either the corresponding gradient vector or the norm of this vector. The Sobel-Feldman operator is based on convolving the image with a small, separable, and integer valued filter in the horizontal and vertical directions. The arrangement of pixels is about the pixel [i, j] shown in **Table 1**. The Sobel's operator is the magnitude of the gradient computed by:

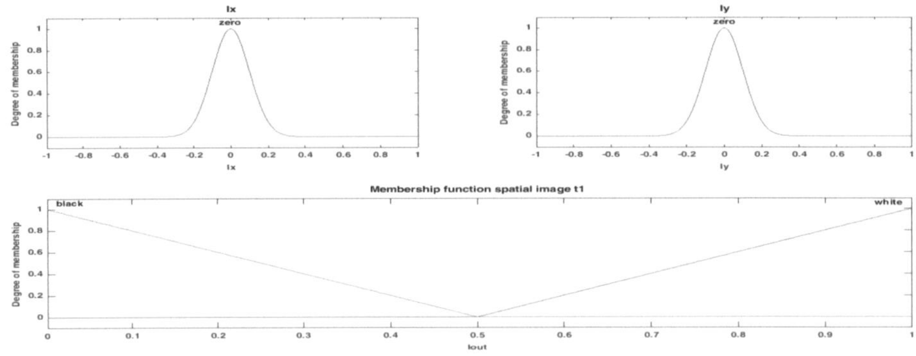


**Figure 7.**
*Define FIS for edge detection (the first and second spatial image). (fuzzy inference system).*

$$S_X = \begin{matrix} -1 & 0 & 1 \\ -2 & 0 & 2 \\ 1 & 0 & 1 \end{matrix}$$

$$S_y = \begin{matrix} 1 & 2 & 1 \\ 0 & 0 & 0 \\ -1 & -2 & -1 \end{matrix}$$

**Table 1.**
*Masks used by Sobel's operator [18].*

$$M\sqrt{S_X^2 + S_Y^2}$$

$$S_X = (a_2 + ca_3 + a_4) - (a_0 + ca_1 + a_6)$$

With the constant $c = 2$.

Like the other gradient operators, Sx and Sy can be implemented using convolution masks:

#### *2.2.1.3.3 Prewitt's operator to edge detection*

Prewitt's operator uses the same equations as Sobel's operator, where constant $c = 1$ (**Table 2**) [19].

### *2.2.1.4 Images comparison with different techniques and error reduction*

#### *2.2.1.4.1 The structural similarity index measure*

The SSIM[2] formula is based on three comparison measurements between the samples, namely the luminance term, the contrast term, and the structural term. The overall index is a multiplicative combination of the three terms [20].

$$\text{SSIM}(x, y) = [l(x, y)]^{\alpha} \cdot [c(x, y)]^{\beta} \cdot [s(x, y)]^{\gamma}, l(x, y) = \frac{2\mu x \mu y + C1}{\mu 2x + \mu 2y + C1},$$

$$c(x, y) = \frac{2\sigma x \sigma y + C2}{\sigma 2x + \sigma 2y + C2}$$

$$s(x, y) = \frac{\sigma xy + C3}{\sigma x \sigma y + C3}$$

$$S_X = \begin{matrix} -1 & 0 & 1 \\ -1 & 0 & 1 \\ -1 & 0 & 1 \end{matrix}$$

$$S_y = \begin{matrix} 1 & 1 & 1 \\ 0 & 0 & 0 \\ -1 & -1 & -1 \end{matrix}$$

**Table 2.**
*Masks used by Prewitt gradient operator.*

2 Structural Similarity Index measure.

Where $\mu x$, $\mu y$, $\sigma x$, $\sigma y$, and $\sigma xy$ are the local means, standard deviations, and cross-covariance for images x, y. If $\alpha = \beta = \gamma = 1$ (the default for Exponents), and $C3 = C2/2$ (default selection of C3) the index simplifies to:

$$\text{SSIM}(x, y) = \frac{(2\mu x\mu y + C1)(2\sigma xy + C2)}{(\mu 2x + \mu 2y + C1)(\sigma 2x + \sigma 2y + C2)}$$

And, dissimilarity Structural is: (1- SSIM(x,y)).

SSIM measures the perceptual difference between two similar images. It cannot judge which of the two is better: that must be inferred from knowing which the "original" and which has been subjected to additional is processing such as data compression.

*2.2.1.4.2 Spatial image subtracts*

Each grayscale image is a matrix with color code (0–256), so we can subtract the two matrices to compare the difference between the two images, here we also used the MATLAB image expression box command, Z = imsubtract (img1, img2).

*2.2.1.4.3 Absolute difference between the two spatial images*

Another way to compare the differences between two spatial images is defined as the sum of the absolute difference at each pixel. The difference value is defined as

$$D(t) = \sum_{i=0}^{M} |I_{(t-T)}(i) - I_t(i)|$$

Where "M" is the resolution or number of pixels in the image. This method for image difference is noisy and extremely sensitive to camera motion and image degradation. When applied to sub-regions of the image, D (t) is less noisy and may be used as a more reliable parameter for image difference.

$$Ds(t) = \sum_{j=s}^{H/n} \sum_{i=s}^{w/n} |I_{(t-T)}(i,\ j) - I_t(i, j)|$$

Ds (t) is the sum of the absolute difference in a sub-region of the image, where S represents the starting position for a particular region, and n represents the number of sub-regions [21]. There is a function to the MATLAB image processing toolbox to compare two images.

$$Z = \text{imabsdiff (img a, img b)}$$

*2.2.1.4.4 Histogram comparison*

In this part of the model, the histogram of two spatial images is drawn and compared in a map that compares the color changes between 0 and 256 from black to white in spatial images.

*2.2.1.4.5 Error calculation and reduction*

To check the error rate in the model, it is necessary to compare two images taken exactly the same in terms of space and time from Google Earth with the model.

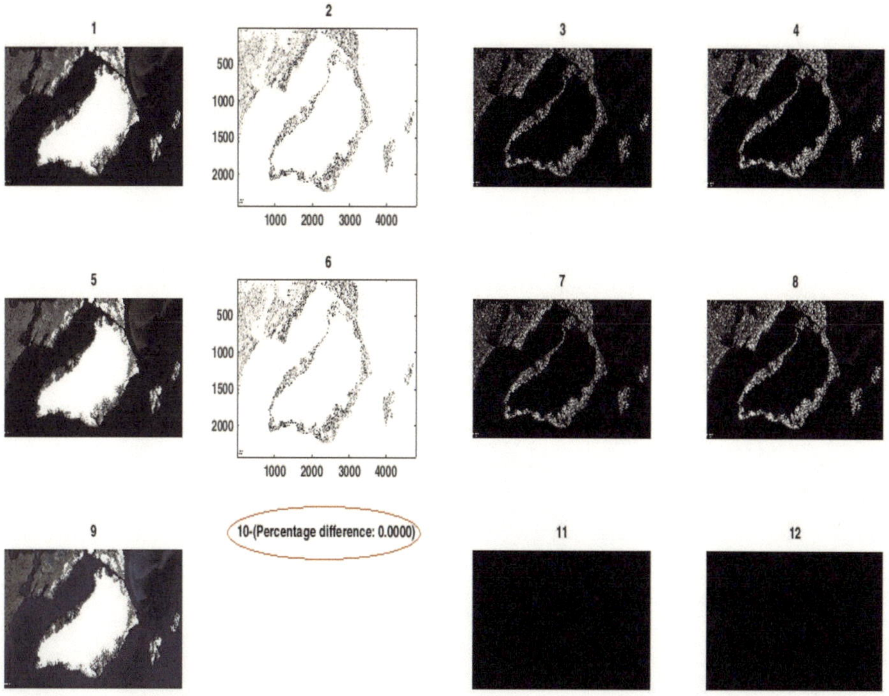


**Figure 8.**
*Calculate the error rate with an accuracy of 4 decimal places, image description according to* ***Table 3****.*

Therefore, ideally, the model should not show any difference. In other words, since the two images taken are exactly the same, the difference calculated by the model must be absolute zero. The accuracy of error calculation is up to 4 decimal places. A comparison of two exactly identical images is shown below (**Figure 8**):

*2.2.1.5 Information integration*

This part, same a single system that information must be shared across all of the functional areas, and the information collected are integrated [22].

*2.2.1.6 Spatial changes recognition*

We use the following method to calculate the difference between the two spatial images.

$$\text{Percentage difference} = (1\text{-SSIM}) * 100\text{–Error}$$

And also, we will provide a complete map to show the spatial changes along with calculating the percentage of changes.

## 3. Result

In this section, after introducing the materials and methods, we run the proposed model to obtain the results. Description of the comparative general map in results is shown in **Table 3**.

| Image number | Description |
|---|---|
| 1 | spatial image time1, RGB to grayscale |
| 2 | Actions: Fuzzy interface system on the spatial image time1 |
| 3 | Actions: Prewitt's operator on the spatial image time1 |
| 4 | Actions: Sobel's operator on the spatial image time1 |
| 5 | spatial image time2, RGB to grayscale |
| 6 | Actions: Fuzzy Interface system on spatial image time2 |
| 7 | Actions: Prewitt's operator on spatial image time2 |
| 8 | Actions: Sobel's operator on spatial image time2 |
| 9 | spatial image time1 (.jpg) format |
| 10 | Difference1, between 6&2, show percentage difference: (1-SSIM) *100 |
| 11 | Defference2, between 7,3 - apply operator imsubtract |
| 12 | Defference3, between 8,4- apply operator imabsdiff |

**Table 3.**
*Description of the comparative general map in results.*

**Experiment 1:**
See **Figures 9–13**.
**Experiment 2:**
See **Figures 14–18**.
**Experiment 3:**
See **Figures 19–23**.
**Experiment 4:**
See **Figures 24–28**.

## 4. Discussion

Spatial analysis in GIS knowledge is done for experts in this field through specialized software such as ArcGIS, etc., and the issue of recognition of environmental

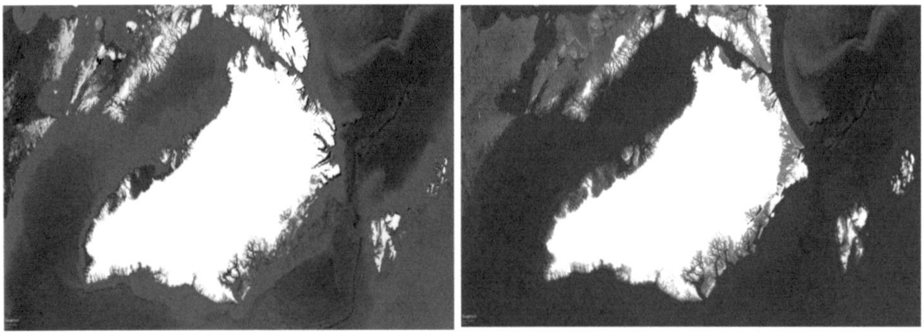

A B

**Figure 9.**
*A: Greenland 1930, B: Greenland2021, input spatial image to model.*

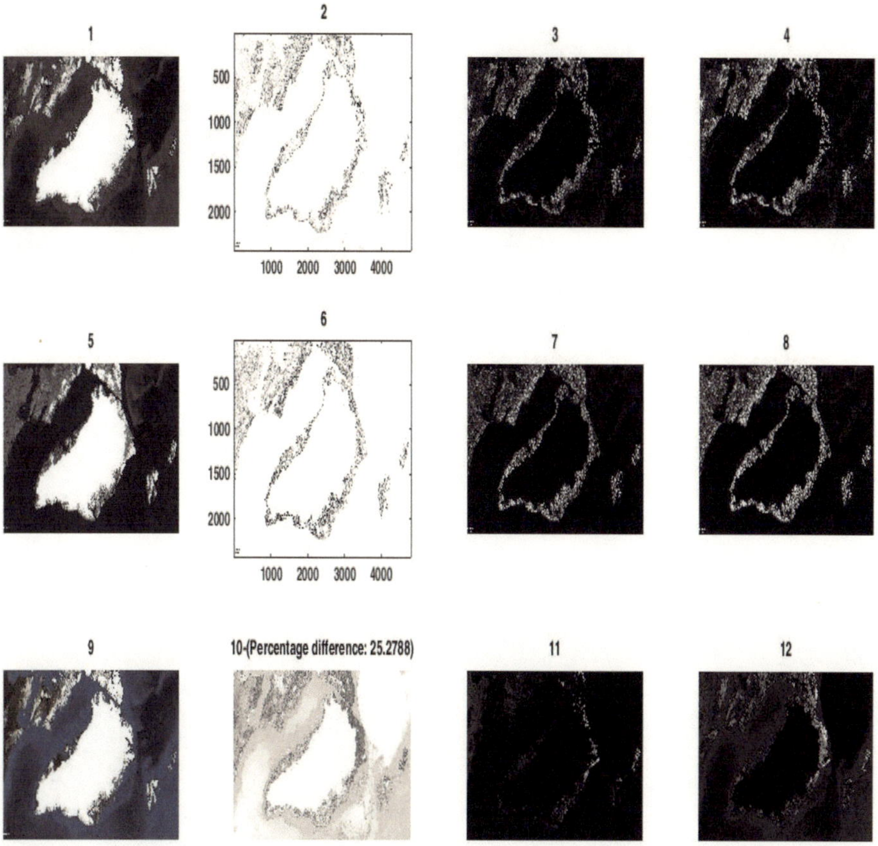


**Figure 10.**
*Comparative general map, the description of the components is as shown in* ***Table 3****.*

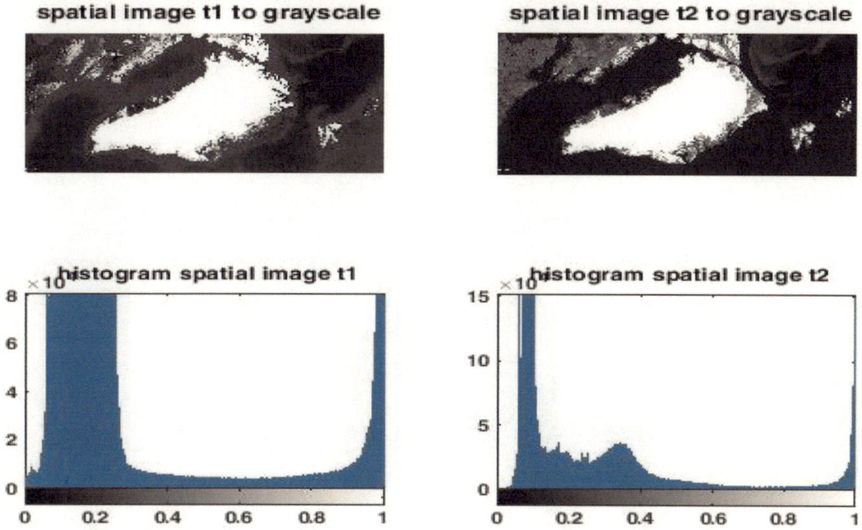


**Figure 11.**
*Comparison of histograms of two temporal spatial images.*

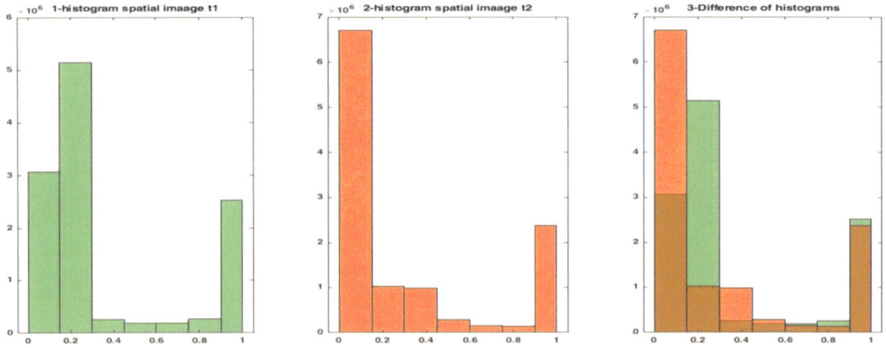


a b c

**Figure 12.**
*a: Bar histogram of spatial image in time 1, b: Bar histogram of spatial image in time 2, c: Histogram comparison.*

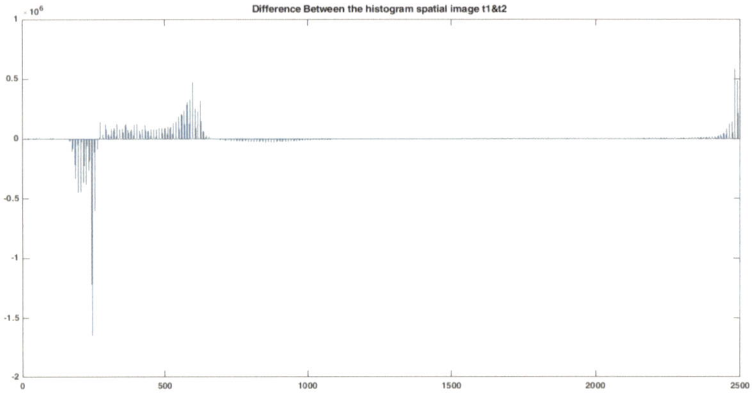


**Figure 13.**
*Histogram compression, above axis: The spatial image in time 2 and below: The spatial image in time 1.*

a b

**Figure 14.**
*a: Jumeirah Palm beach 2000, b: Jumeirah Palm beach 2000.*

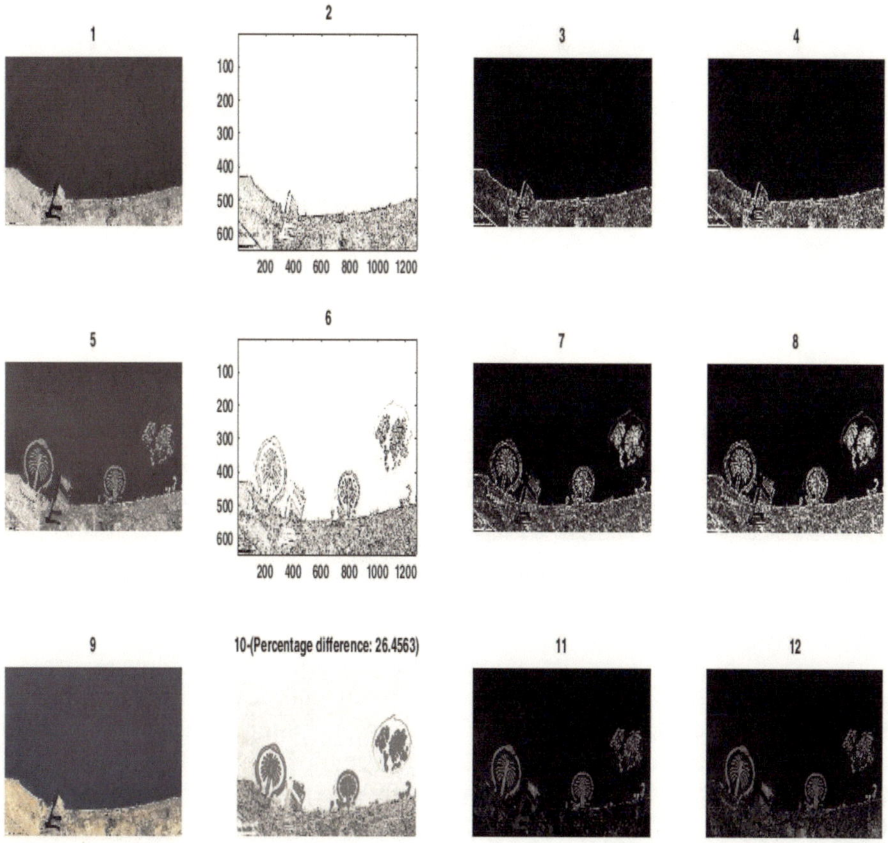


**Figure 15.**
*Comparative general map, the description of the components is as shown in **Table 3**.*

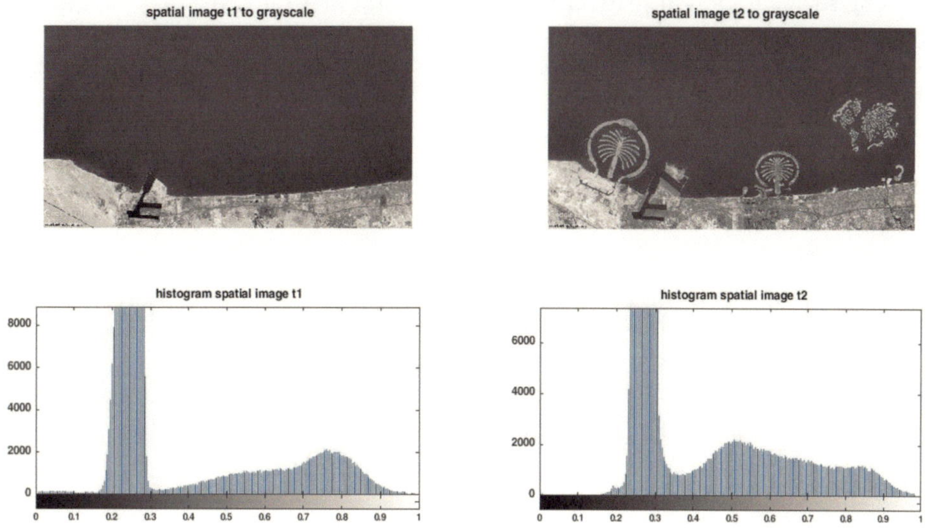


**Figure 16.**
*Comparison of histograms of two temporal spatial images.*

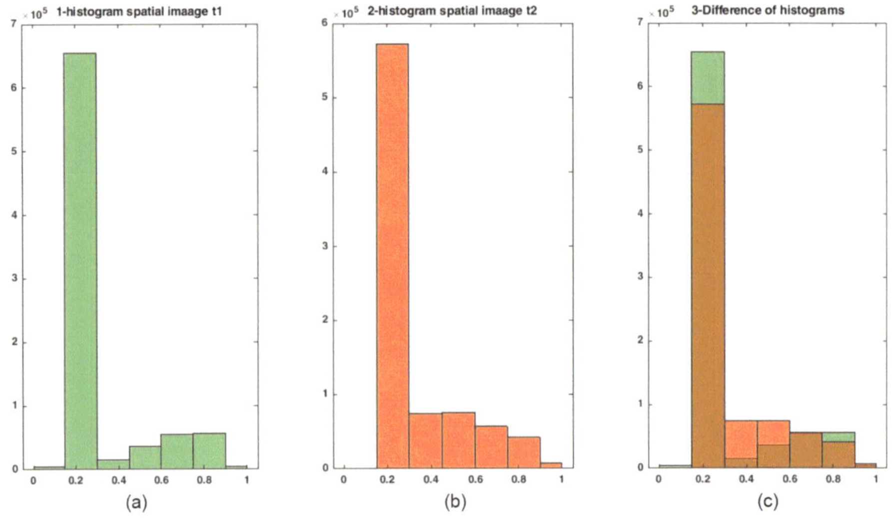


**Figure 17.**
*a: Bar histogram of spatial image in time 1, b: Bar histogram of spatial image in time 2, c: Histogram comparison.*

changes, etc. is of special importance to them. In the preparation of image data according to the rules of data mining, the only change in the captured images should be time and other variables of the desired location, including geographical characteristics, camera height from the ground, etc. must remain constant, and otherwise, a calculation error will occur. Therefore, by strictly observing this important point, the system error will be zero; in other words, we will have an accuracy of %100. Also, the test results will be the same no matter how many (repetitions) are performed, which explains the validation of the model.

Comparison segmentation method and proposed model, advantages and disadvantages:

Using segmentation in spatial images is important task for detecting changes. Segmentation must not allow regions of the image to overlap. Thresholding is one of the oldest methods used for image segmentation. It is based on the gray level intensity

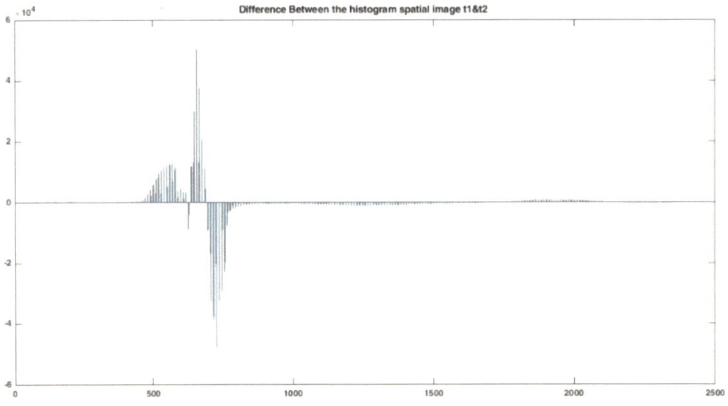


**Figure 18.**
*Histogram compression, above axis: the spatial image in time 2 and below: the spatial image in time 1.*

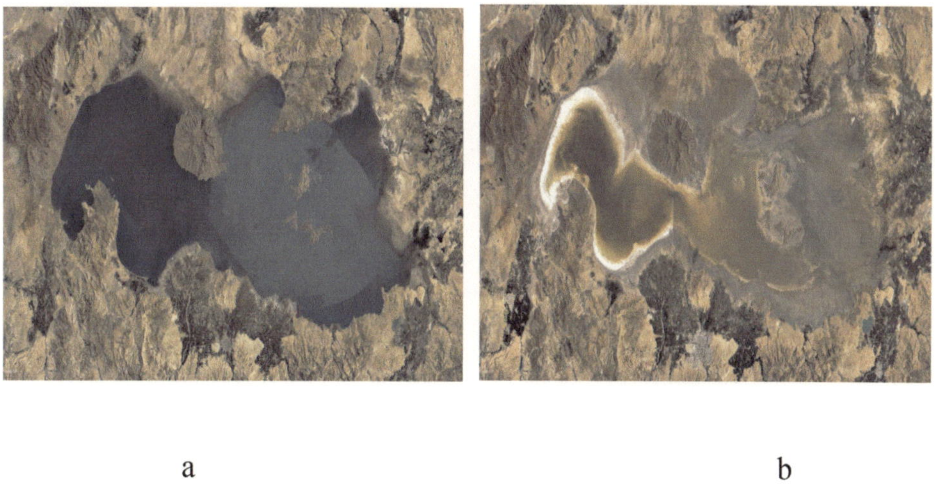


**Figure 19.**
*a: Oroomiye Lake1984, b: Oroomiye Lake 2017.*

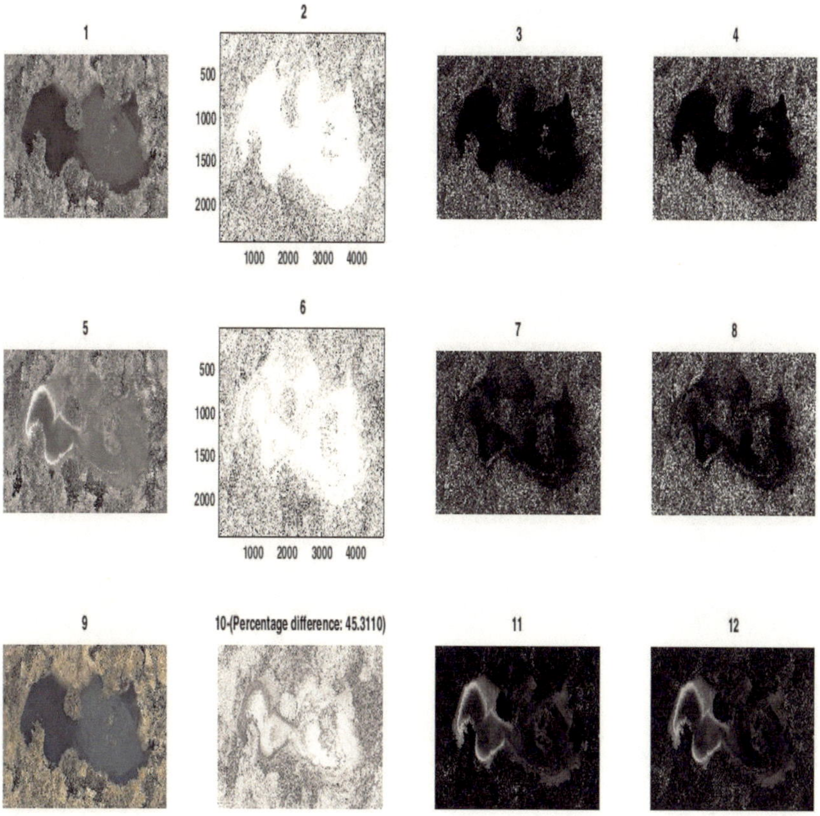


**Figure 20.**
*Comparative general map, the description of the components is as shown in* ***Table 3****.*

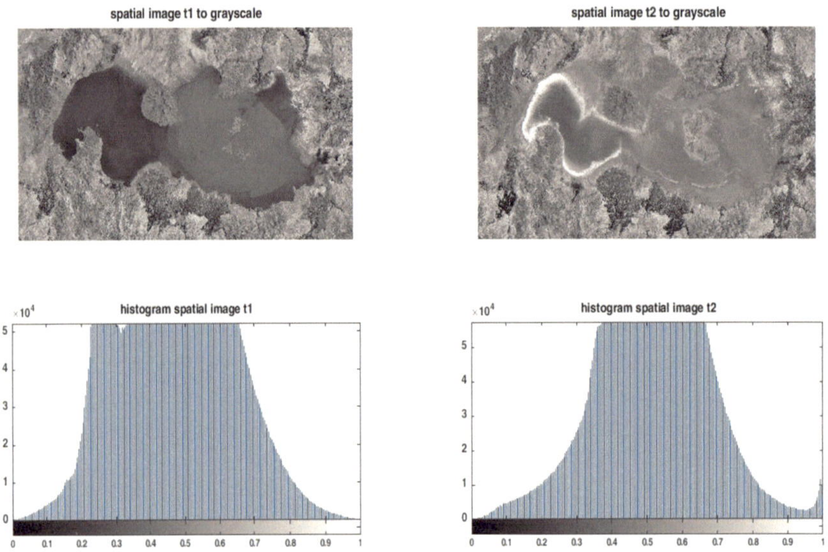


**Figure 21.**
*Comparison of histograms of two temporal spatial images.*

value of pixels. The histogram of an image consists of Conceptually is similar to the classifiers except that they are implemented in the spatial domain of an image rather than in a feature space. It treats the segmentation as a registration process. Some researchers used atlases not only to impose spatial constraints but also to provide probabilistic information about the tissue model. The advantage is that it can segment an image with no well-defined relation between regions and pixels. K-means is a clustering method that partitions the n-points into the k-clusters in which each pixel belongs to one cluster by minimizing an objective function in such a way that within a cluster sum of squares is get minimized. It starts with k-clusters and each pixel is

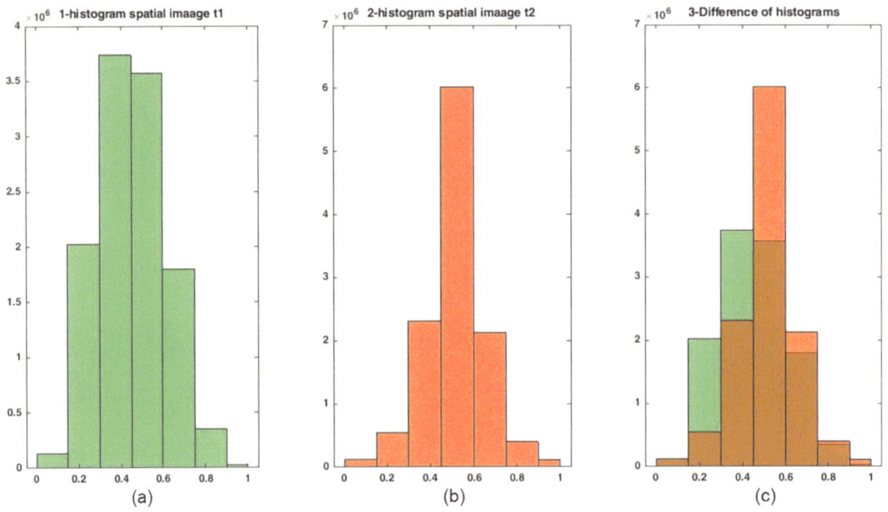


**Figure 22.**
*a: Bar histogram of spatial image in time 1, b: Bar histogram of spatial image in time 2, c: Histogram comparison.*

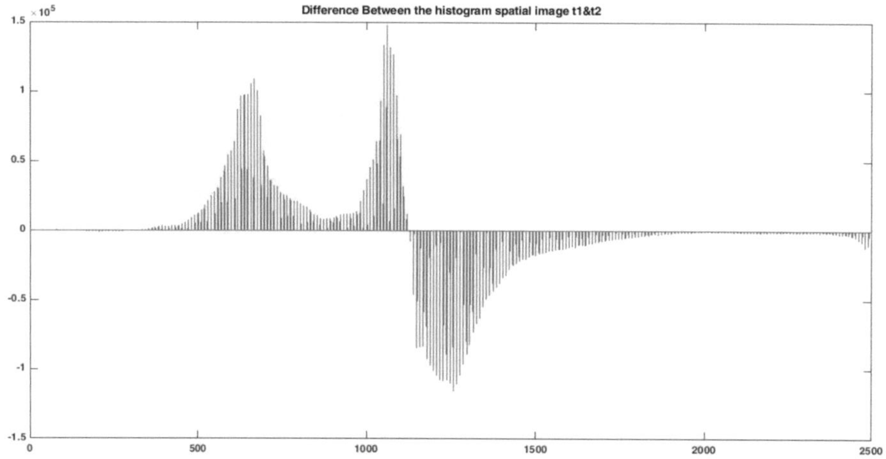


**Figure 23.**
*Histogram compression, above axis: the spatial image in time 2 and below: the spatial image in time 1.*

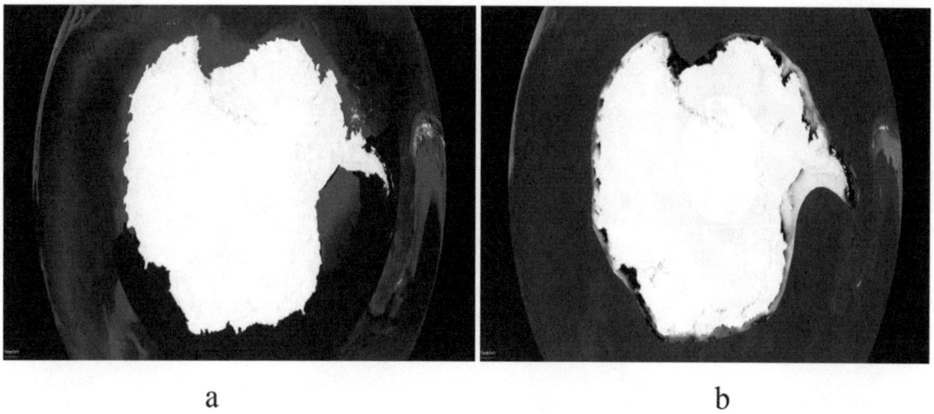

**Figure 24.**
*a: South Pole 1957, b: South Pole 2022.*

assigned to one cluster. The limitation of the K-means algorithm is computational time increases on implementation in large amounts of data but our proposed model is independent of clustering so will be faster than the K-means algorithm.

## 5. Implication

In the proposed model, using machine vision, image data processing, fuzzy mathematical techniques, the use of known masks, as well as historical images in Google Earth software, we can measure changes in images and measure them. This software can be added to Google Earth software as a development part and users can easily view spatial changes in terms of time changes.

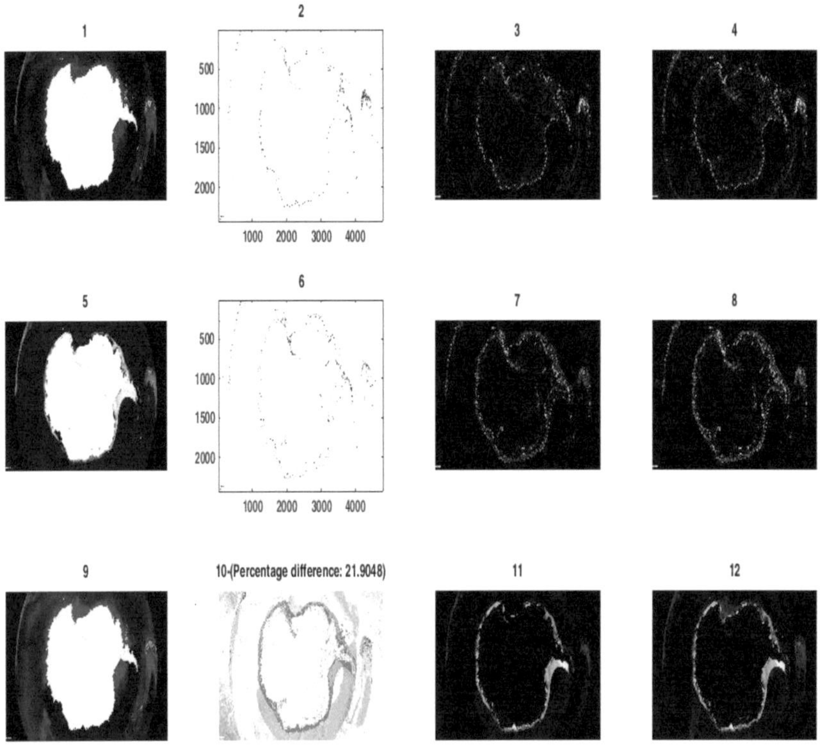


**Figure 25.**
*Comparative general map, the description of the components is as shown in* ***Table*** *3.*

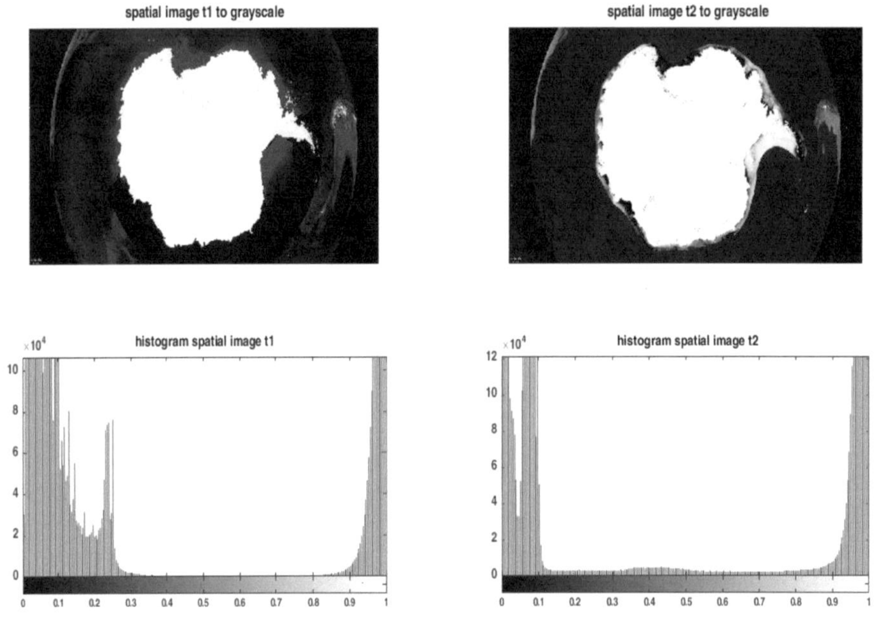


**Figure 26.**
*Comparison of histograms of two temporal spatial images.*

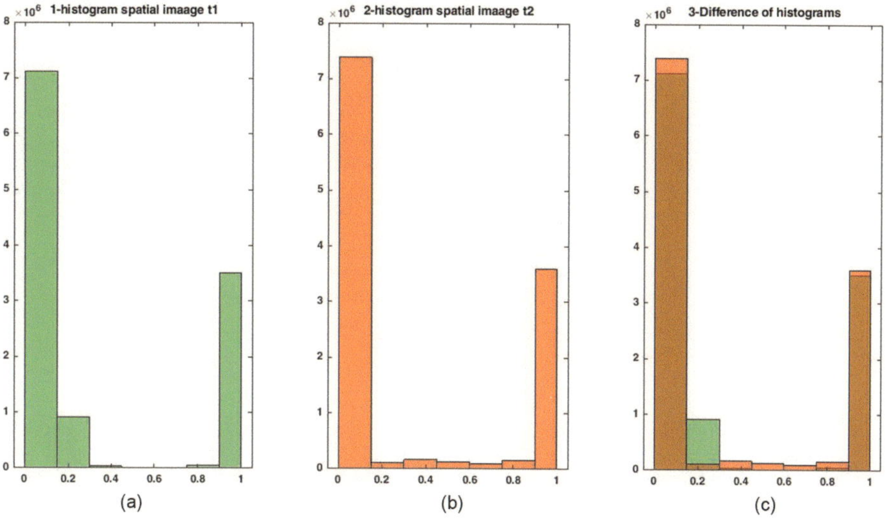


**Figure 27.**
*a: Bar histogram of spatial image in time 1, b: Bar histogram of spatial image in time 2, c: Histogram comparison.*

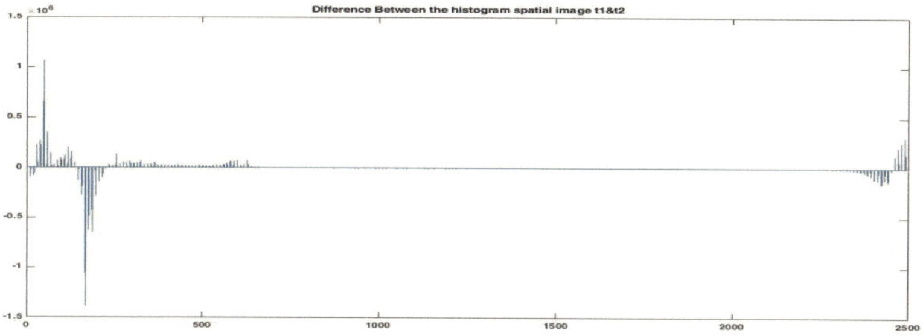


**Figure 28.**
*Histogram compression, above axis: the spatial image in time 2 and below: the spatial image in time 1.*

## 6. Conclusion

The change recognition model presented by the author of the article earlier [17], was also used in spatial change recognition. The most important difference between the previous models in change recognition in industrial parts with the current model is in the field of capturing images and image preparation. Here the images are taken from Google Earth and no filter is used to prepare them because the images are of sufficient quality and adding any filter will cause a computational error in the model (this was tested many times by the authors) but in the base model, the capture of images was with a local camera that had the error and the macrographic imaging technique was done [17].

## Author details

Majid Mirbod
Department of Industrial Management, Tehran North Branch, Islamic Azad University, Tehran, Iran

*Address all correspondence to: mjmirbod@yahoo.com

## References

[1] Khurana M, Saxena V. Soft computing techniques for change detection in remotely sensed images: A review. International Journal of Computer Science Issues. 2015;**12**(2)

[2] Bruzzone L, Bovolo F. A novel framework for the Design of Change-Detection Systems for very-high-resolution remote sensing images. IEEE. 2013;**101**(3)

[3] Inzamul Haque M, Basak R. Land cover change detection using GIS and remote sensing techniques: A spatio-temporal study on Tanguar Haor, Sunamganj, Bangladesh. The Egyptian Journal of Remote Sensing and Space Sciences. 2017;**20**(2):251-263. DOI: 10.1016/j.ejrs.2016.12.003

[4] Gomaa M, Hamza E, Elhifnawy H. Automated unsupervised change detection technique from RGB color image. Materials Science and Engineering. 2019;**610**:012046. DOI: 10.1088/1757-899X/610/1/012046

[5] Fisher R. Change detection in color images. In: Proceedings of 7th IEEE Conference on Computer Vision and Pattern. Mathematics. Citeseer. 1999

[6] Kadhum ZM, Jasim BS, Obaid MK. Change detection in city of Hilla during period of 2007-2015 using remote sensing techniques. Materials Science and Engineering. 2020;**737**:012228. DOI: 10.1088/1757-899X/737/1/012228

[7] Wales N, Murphy RJ, Bruce E. Understanding patterns of vegetation change at the Angkor world heritage site by combining remote sensing results with local knowledge. International Journal of Remote Sensing. 2021;**42**(2). DOI: 10.1080/01431161.2020.1809739

[8] Bouziani M, Goïta K, He D-C. Automatic change detection of buildings in urban environment from very high spatial resolution images using existing geodatabase and prior knowledge. ISPRS Journal of Photogrammetry and Remote Sensing. 2010;**65**:143-153. DOI: 10.1016/j.isprsjprs.2009.10.002

[9] You Y, Cao J, Zhou W. A survey of change detection methods based on remote sensing images for multi-source and multi-objective scenarios. Remote Sensing. 2020;**12**(15):2460. DOI: 10.3390/rs12152460

[10] Li Q, Wang S, Bai X, Luo G, Song X, Tian Y, et al. Change detection of soil formation rate in space and time based on multi source data and geospatial analysis techniques. Remote Sensing. 2020;**12**:121. DOI: 10.3390/rs12010121

[11] Panuju DR, Paull DJ, Gri AL. Change detection techniques based on multispectral images for investigating land cover dynamics. Remote Sensing. 2020;**12**:1781. DOI: 10.3390/rs12111781

[12] Asokan A, Anitha J. Change detection techniques for remote sensing applications: A survey. Earth Science Informatics. 2019;**12**:143-160. DOI: 10.1007/s12145-019-00380-5

[13] Kirscht M, Rinke C. 3D reconstruction of buildings and vegetation from synthetic aperture radar (SAR) images. MVA. 1998

[14] Lia Z, Jia Z, Liu L, Yang J, Kasabovc N. A method to improve the accuracy of SAR image change detection by using an image enhancement method. ISPRS Journal of Photogrammetry and Remote Sensing. 2020;**163**:137-151. ISSN: 0924-2716. DOI: 10.1016/j.isprsjprs.2020.03.002

[15] Sun Y, Lei L, Li X, Sun H, Kuang G. Nonlocal patch similarity based heterogeneous remote sensing change detection. DOI: 10.1016/j.patcog.2020.107598

[16] Ansari RA, Buddhiraju KM, Malhotra R. Urban change detection analysis utilizing multiresolution texture features from polarimetric SAR images. Remote Sensing Applications: Society and Environment. DOI: 10.1016/j.rsase.2020.100418

[17] Mirbod M, Ghatari AR, Saati S, Shoar M. Industrial parts change recognition model using machine vision, image processing in the framework of industrial information integration. Journal of Industrial Information Integration. 2022;**26**:100277. DOI: 10.1016/j.jii.2021.100277. ISSN: 2452-414X

[18] Kanopoulos N et al. Design of an Image Edge Detection Filter using the Sobel operator. Journal of Solid-State Circuits, IEEE. 1988;**23**(2):358-367

[19] Seif A et al. A hardware architecture of Prewitt edge detection. In: Sustainable Utilization and Development in Engineering and Technology (STUDENT), 2010 IEEE Conference. Computer Science. Malaysia; 2010. pp. 99-101

[20] Zhou W, Bovik AC, Sheikh HR, Simoncelli EP. Image quality assessment: From error visibility to structural similarity. IEEE Transactions on Image Processing. 2004;**13**(4):600-612

[21] Xiong Z, Huang TS. The Essential Guide to Video Processing. Texas, USA: Department of Electrical and Computer Engineering The University of Texas at Astin; 2009

[22] Xu L. Enterprise Integration and Information Architecture: A Systems Perspective on Industrial Information Integration. Auerbach Publications; 2014. p. 446. ISBN: 9781439850244

Chapter 8

# Methods for Real-time Emotional Gait Data Collection Induced by Smart Glasses in a Non-straight Walking Path

*Nitchan Jianwattanapaisarn, Kaoru Sumi and Akira Utsumi*

**Abstract**

Emotion recognition is an attractive research field because of its usefulness. Most methods for detecting and analyzing emotions depend on facial features so the close-up facial information is required. Unfortunately, high-resolution facial information is difficult to be captured from a standard security camera. Unlike facial features, gaits and postures can be obtained noninvasively from a distance. We proposed a method to collect emotional gait data with real-time emotion induction. Two gait datasets consisting of total 72 participants were collected. Each participant walked in circular pattern while watching emotion induction videos shown on Microsoft HoloLens 2 smart glasses. OptiTrack motion capturing system was used to capture the participants' gaits and postures. Effectiveness of emotion induction was evaluated using self-reported emotion questionnaire. In our second dataset, additional information of each subject such as dominant hand, dominant foot, and dominant brain side was also collected. These data can be used for further analyses. To the best of our knowledge, emotion induction method shows the videos to subjects while walking has never been used in other studies. Our proposed method and dataset have the potential to advance the research field about emotional recognition and analysis, which can be used in real-world applications.

**Keywords:** emotion induction, emotion recognition, gait analysis, motion capturing, smart glasses, non-straight walking behavior, emotional movies, watching video while walking

## 1. Introduction

Intelligent video surveillance research gains a lot of interests by the public. In this study, an example of applications was conducted to show the potential of monitoring human behaviors from their movements. The authors have conducted some studies to analyze the characteristics of individuals. In order to conduct research about recognition of human emotions, the authors proposed a research environment and research method in which human emotions can be changed in real time using video stimuli, and experiments on emotion recognition can be performed using our proposed environment and method. Recognizing human emotions is very useful in several circumstances, for example, improvement of human-robot interaction experiences, suspicious behaviors detection for crime and altercation prevention, customer satisfaction evaluation, students' engagement evaluation. These are some examples of applications that can improve the quality of life for humans.

Affective computing [1] is a specific research field that was emerged because of the popularity of emotion analysis research, which attempts to make a computer to be able to understand and generate human-like affects. There are many studies related to affective computing proposed during recent years. A good example is about the online exercise program for students to practice their programming skills. Affective computing technique was applied to an online exercise program by analyzing the emotion of students as well as their performance in each task. Then, an animated agent is used to interact with students during the exercise so the students can interact with the agent. This method can improve students' experiences and performances at the same time [2].

Another good example is about surveillance and security application, which is related to intelligent video surveillance topic of this book. A survey conducted by [3] reveals that gait analysis is very useful for crime prevention applications. In recent years, a CCTV camera system is now a standard equipment, which is installed in almost every public places. Human gaits can be analyzed in a very short time due to the advancement of computer vision and machine learning technology together with the help of on-board computation devices. Therefore, suspicious behaviors can be detected promptly. According to a study [4], smart video surveillance is very useful for many applications by applying gait analysis techniques such as human identification, human re-identification, and forensic analysis because human gaits can be obtained from far away without the subjects' awareness or cooperation.

In emotion recognition and prediction aspect, in the past, these applications were performed by human observers [5]. Unfortunately, using human observer to judge the emotions of other subjects is time consuming. Humans as the judges are not consistent enough to be used in reality. As a result, automatic emotion methods have been developed. Most publicly available methods for emotion recognition nowadays are performed using facial expression. Facial features perform very well in some situations. However, there are still some limitations of using facial features to perform emotion recognition. If the facial images or videos must be captured in a crowded and noisy environment, it is difficult to capture high-quality facial features since a standard security camera cannot perform well enough. Particularly, when the subject is not facing forward with the camera, facial features are difficult to be obtained accurately. Moreover, some subjects wear eyeglasses and sunglasses, or have beard and mustache, which also prevent emotion analysis from facial features to be performed effectively. Therefore, if facial features cannot be clearly captured, other features should be used instead to make emotion recognition and analysis more practical for real-world uses.

Gaits and postures are the way that human body moves and poses while they are walking. This kind of expressions can be observed from distance without awareness of subjects. Also, it can be captured without the need for high-resolution images or videos. Thus, gaits and postures are very good expressions of a human that can be used for emotion recognition and analysis. There are several applications that can be performed effectively and accurately using human gaits and postures such as human identification [6, 7] human re-identification [8], human age estimation, and gender recognition [9, 10]. From many studies proposed, these prove that human gait and posture are very appropriate features for prediction and recognition of human emotions [5, 11–20].

The objective of this study is to propose a method for emotional gait data collection using a novel method to induce subjects' emotions when the subjects are walking in a non-straight walking direction, which is an unconventional walking path and cannot be found in other studies. This study proposed a method and environment to perform gait data collection in different emotions. Microsoft HoloLens 2 smart glasses were used for displaying the emotion induction videos to the participants while they are walking. OptiTrack motion capturing system was used to record the walking data of subjects. Although we used the OptiTrack, which is a marker-based devices, to record body movements while walking, other marker-less motion capture devices such as Microsoft Kinect or Intel RealSense can be also used instead of OptiTrack. Also, because of the advancement of pose-estimation software, for example, OpenPose or wrnch AI, any video cameras can be used to capture human gaits for gait analysis.

## 2. Related work

Many studies on emotion recognition were proposed in recent years because of their usefulness. Most of them were performed using facial features, which are sufficiently accurate in some situations. However, facial features still have some limitations as discussed in the previous section. Even though there are many studies that used human gaits and postures as features for emotion recognition, the number is still fewer than the studies using facial features.

A survey conducted by [21] investigated several studies about gait analysis, not only for emotion recognition but also for human identification. They found that characteristics of human walking are different in different emotions. This information can be used for development of automatic emotion recognition. In comparison with other biometrics, for example, speech features, facial features, physiological features, they found that using gaits has many advantages. For instance, gaits can be observed without subject's awareness from afar, imitation of gaits are very difficult, and subject's cooperation is not required to obtain human gaits. Hence, gaits are very powerful expressions, which can be used to perform automatic emotion recognition. We are going to mention only about the equipment that can be used for collection of gait data and the results, which shows the effectiveness of emotion prediction from gaits. From this survey study, several devices can be used to capture gait data. For instance, a force plate can be used to record velocity and pressure data [11], velocity data can be recorded well by using an infrared light barrier system [11, 22], motion capturing devices such as Vicon are a good tool to record the coordinates of body parts by attaching the markers on subjects' body [12–15, 23–25], wearable devices such as smart watches equipped with accelerometer as well as smart phones can be used to record body movements data to use for gait analysis [18–20], and Microsoft Kinect is also an effective tool for recording human skeleton without the need of markers to be attached on subjects' body [6–9, 16, 17, 26]. Some findings are useful for future studies are as follows. When the subjects feel happy, they step faster [5] and their strides are longer [27]. Also, their joint angles amplitude [27] and arm movements [12] increase.

When the subjects feel sad, their arm swings decrease [5], and their limb shape and torso shape [24] are contracted. Their joint amplitude also decreases [12].

Nowadays, there are several studies about gait analysis proposed. Some examples include emotion prediction [11, 21], mental illness prediction [22, 23], human identification or re-identification [6–8], and gender prediction [9, 10]. Several tools can be used for gait data collected as already mentioned, for example, light barrier, force plate, video camera, accelerometer, motion capturing system. From this equipment, we focus on the equipment that captures the coordinates of body parts or silhouette images of human body since these gait features are sensitive to walking directions. Using straight walking direction usually results in high-quality gait data [11, 12, 14–17, 19, 22, 23, 26, 28–30], so most studies used this type of walking direction. There are fewer works that used free-style walking pattern; that is, subjects can choose the walking path any direction the want [6–9]. By using free-style walking data, the results often lower than using straight walking data, but it increases the opportunities to deploy the proposed methods in reality. Since human walking in public spaces is always lack of awareness for being observed and the walking pattern cannot be controlled to be a straight walking path. That is, to collect the straight walking gait data in real-world environment is more difficult than random direction walking data.

In this study, we decided to use the latest technology smart glasses called Microsoft HoloLens 2 to display the emotional videos to the subjects while they are walking. Therefore, we concerned some issues including the interference of smart glasses to human gaits while walking, negative effects such as trips and slips are also important to be considered. Some studies were performed on this topic, and they are useful for our study. For example, a study proposed by [31] performed an investigation of gait performance while the subjects use head-worn display during walking. Experiment was done using 12 subjects to check whether the subjects can walk normally in different conditions. Several factors were assessed, that is, walking speed and obstacle crossing speed, required coefficient of friction, minimum foot clearance, and foot placement location around the obstacle. From this study, they found that using head-worn display to perform tasks while walking has no effect with level walking performance when comparing with using a paper list and with baseline walking that used nothing. For obstacle crossing experiment, they found that the subjects choose more cautious and more conservative strategy to cross the obstacle if they are using the head-worn display. Obstacle crossing speed also decreases by 3% when compared with the baseline walking. Besides, using head-worn display does not affect with foot placement location around the obstacle.

Other useful studies that investigated the negative effects on human gaits when using head-worn display are [32, 33]. They performed experiments to find out the adverse effect when the subjects use head-worn display while walking. They asked 20 subjects (10 men and 10 women) to walk in four different conditions on a treadmill. Subjects were asked to perform one single-task walk (walking and do nothing) and three dual-task walks (walking and perform attention-demanding tasks). Dual-task walks were conducted by different display types including paper-based, smart phone, and smart glasses for displaying information to the subjects while they walk. Attention-demanding tasks include Stroop test, categorizing, and arithmetic. The subjects use head-down posture while they performed tasks on paper-based display and on smart phone. In single-task walking and in dual-task walking using smart glasses, they use head-up posture. Vicon motion capture system with seven cameras was used in their experiments. The results of their studies reveal that walking while

*DOI: http://dx.doi.org/10.5772/intechopen.107410*

using smart glasses to perform attention-demanding tasks has more impacts with gait performance, for example, gait stability in comparison with walking while performing attention-demanding tasks on other display types. The important finding from their studies is that the subjects are more unstable if they use smart phone and paper-based display to perform tasks while walking than using smart glasses. This means that the head-up and head-down postures affect with human gaits.

From reviewing of related works, Microsoft HoloLens 2 was confirmed that it can be used for displaying videos as the subjects can use head-up posture while walking, and they can also see the room environment while watching videos since the HoloLens 2 display is transparent. Even though there could be some negative impacts such as walking stability or obstacle crossing strategy, we cope with these issues by asking our subjects to take rehearsal walks to make them familiar with using HoloLens while walking and with the walking area before performing the actual recording walks. About the obstacle, our walking space is very clear so there should be no problem with using HoloLens 2 while walking.

## 3. Data collection

Gait data collection method described in this study has been proposed by us [34]. The data collection method we used is as follows. Since most studies in emotion recognition and analysis using gaits and postures were performed by asking subjects to walk straightly on the pathway or on the treadmill, we found that walking in a straight line will result in cleaner gait data. However, it will be more difficult to be implemented in reality. In emotion induction aspect, there are some techniques widely used as follows. First, subjects are asked to walk while recalling their own personal experiences according to assigned emotions. Second, subjects are not normal people but professional actors. Third, subjects are asked to watch an emotional video on a conventional screen such as television or computer display before they start walking.

With these settings, it is possible that some problems can occur. In the first method, subjects may not be able to recall their memories well enough to express the desired emotions on their gaits and body movements. In the second method, using professional actors instead of normal people can make the gaits too exaggerate and not natural. In the last method, it is possible that the induced emotions will not last until the end of walking because the video stimuli end before the subjects start walking. These issues can make the collected gait data incorrectly reflect human emotions and the relationship between collected gaits and emotions will be inaccurate.

In order to solve this problem, our experiments were designed to make the subjects watch a video for emotion induction and walk at the same time to record real-time emotion of subjects. Since there is a latest smart glasses technology named Microsoft HoloLens 2 available for consumer uses, we decided to use HoloLens 2 for displaying emotional videos to the subjects while they are walking. With this method, subjects can watch the stimuli at the same time they walk; hence, their emotions will be constantly and consistently induced. As of now, to the best of our knowledge, there is no other researcher who used this method before. Because of the transparent display of HoloLens 2, subjects can see the walking space and the room environment at the same time while they walk. We also expected that showing videos to the subjects during walking will be more similar to when subjects walk in real life and see some situations that make their emotion change in real time according to those situations. This emotion induction method is planned to simulate the subject's real-time emotion. Also, because we showed the videos at the same time of walking, we can ensure that the induced emotions will be more stable, more consistent, and last until the end of the walk.

For the walking direction, because our subjects have to watch the videos for emotion induction and walk in the walking space at the same time, allowing them to walk freely without path guidance at all can be too difficult for them. Because subjects need to concentrate with the content of the videos, if they also have to select the walking path while walking, it is possible that they will not be able to focus on the videos well enough and the emotion induction will not be effective. Consequently, we asked the subjects to walk in a circular pattern without guidance line on the floor. That is, subjects can walk in lax circular path, clockwise, or counter-clockwise depending on their own preferences. With this walking direction, subjects can walk like an oval shape or like a rounded-rectangle shape as they want. Therefore, we can collect both straight walking and non-straight walking in one walking trial.

### 3.1 Equipment for data collection

Motion capturing devices can be categorized into two main types. First, marker-less type, which is easy to setup and requires nothing to be attached on subject's body. Second, marker-based type, which requires several markers to be attached on subject's body. Differences between these two types are marker-less type uses image processing and machine learning technology to predict the positions of body parts from depth image and color image captured from build-in cameras, while marker-based type requires several cameras to be installed, and the actual position of each marker is calculated from the reflection of infrared light captured by all cameras. This means that the position of each marker is reconstructed from all cameras data to obtain a coordinate in three-dimensional space. This makes the marker-based motion capturing device more accurate but also more difficult to setup, whereas marker-less device such as Microsoft Kinect is much easier to setup and use in any situation.

In this study, we decided to use OptiTrack, which is a famous marker-based motion capturing system to capture human gaits. Fourteen OptiTrack Flex 3 cameras were installed around the recording space, and OptiTrack Entertainment Baseline Markerset consisting of 37 markers was used in our experiments. **Table 1** lists all marker names, and **Figure 1** shows the position of each marker on human body.

### 3.2 Recording environment

The black tape was used for marking a rectangle walking area on the floor as shown in **Figure 2**. Inside the rectangle is the area that OptiTrack can capture. The size of this walking space is 2.9 by 3.64 meters. Fourteen OptiTrack Flex 3 cameras were installed on seven camera stands, and each stand was placed around the walking space as illustrated in **Figure 3**. In other words, one camera stand has two OptiTrack Flex 3 cameras installed at different height levels, one on higher level and another on the lower level as shown in **Figure 4**.

### 3.3 Materials for data collection

Three videos were selected as stimuli to induce subject's emotion. HoloLens 2 was used for displaying these videos to each subject while he or she is walking circularly in the recording area.

*DOI: http://dx.doi.org/10.5772/intechopen.107410*

| | |
|---|---|
| **HeadTop** | |
| **HeadFront** | |
| **HeadSide** | |
| **BackTop** | |
| **Chest** | |
| Back | Left |
| | Right |
| WaistFront | Left |
| | Right |
| WaistBack | Left |
| | Right |
| ShoulderBack | Left |
| | Right |
| ShoulderTop | Left |
| | Right |
| ElbowOut | Left |
| | Right |
| UpperArmHigh | Left |
| | Right |
| WristOut | Left |
| | Right |
| WristIn | Left |
| | Right |
| HandOut | Left |
| | Right |
| ThightFront | Left |
| | Right |
| KneeOut | Left |
| | Right |
| Shin | Left |
| | Right |
| AnkleOut | Left |
| | Right |
| ToeOut | Left |
| | Right |
| ToeIn | Left |
| | Right |

**Table 1.**
*List of OptiTrack baseline markers.*

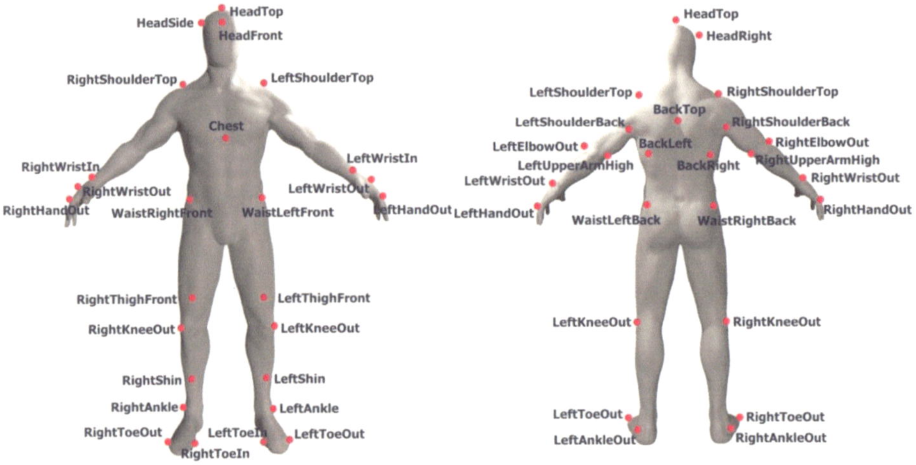


**Figure 1.**
*Position of each marker on the body (Human Figure Source: https://sketchfab.com/3d-models/man-5ae6bd 9271ac4ee4905b96e5458f435d).*

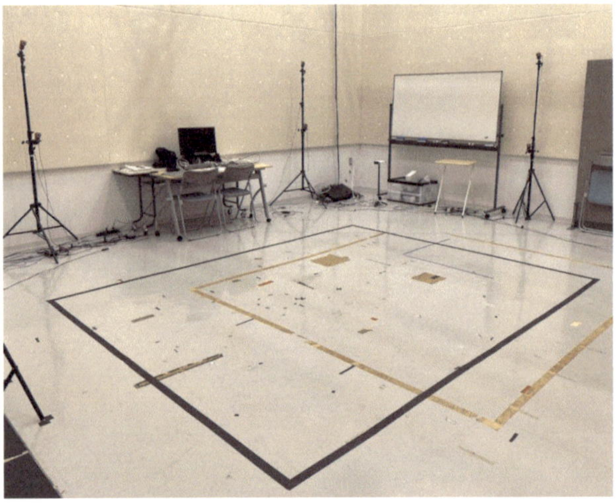

**Figure 2.**
*Rectangle walking area marked with black tape on the floor.*

- **Neutral Video:** The nature landscape video from YouTube named *Spectacular drone shots of Iowa corn fields* uploaded by the YouTube user named *The American Bazaar* (https://www.youtube.com/watch?v=4R9HpESkor8)

- **Negative Video:** An emotional movie selected from LIRIS-ACCEDE database named *Parafundit* by *Riccardo Melato*

- **Positive Video:** An emotional movie selected from LIRIS-ACCEDE database named *Tears of steel* by *Ian Hubert* and *Ton Roosendaal*

The Neutral video was selected from nature landscape videos on YouTube that should not induce any emotion. Positive video (for inducting happy emotion) and

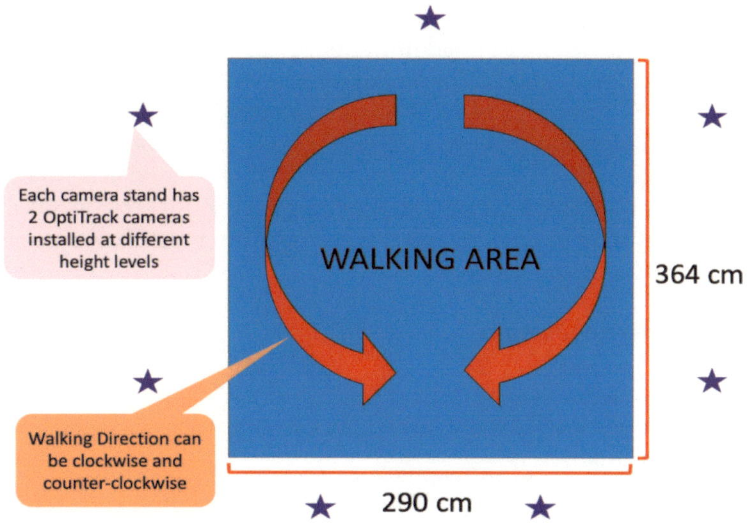


**Figure 3.**
*Position of each camera and dimension of the walking area.*

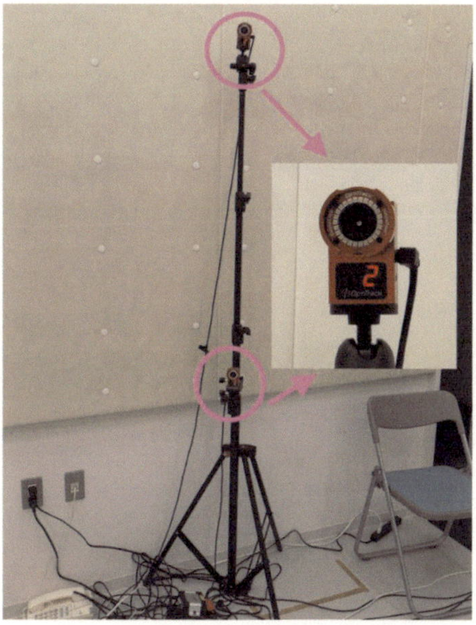

**Figure 4.**
*Two OptiTrack Flex 3 cameras installed on each camera stand at different height levels.*

negative video (for inducting sad emotion) were selected from a public annotate movie database named *LIRIS-ACCEDE* (https://liris-accede.ec-lyon.fr/). This database was published by [35]. It consists of several movies and their emotion annotations in valence-arousal dimension. All movies in this database are published under the Creative Commons license. In our study, we selected two movies from the *Continuous LIRIS-ACCEDE collection*. We found that most movies contain both positive valence and negative valence. As we would like to make an entire walking trial to contain only one emotion, we selected one movie with only positive valence and another movie with only negative valence annotation. As each subject needs to walk when the movie starts until the movie ends, all movies we selected must not be too long. In our opinion, less than 15 minutes in length is acceptable. The lengths of the neutral video, negative movie, and positive movie are 5:04, 13:10, and 12:14 minutes, respectively. Sample plots of valence score for annotated movies are shown in **Figure 5**, and plots of the negative and positive movie we used are shown in **Figure 6**. Neutral video has no sound at all to ensure that it will not induce any emotion. Positive video and negative video contain music, sound effects, and conversation in English. Subjects can hear the audio from stereo speakers, which are build-in with the HoloLens 2.

### 3.4 Methods for data collection

Before participating in our experiments, we kindly asked our participants to answer the health questionnaire and signed the consent form. Questions in the health questionnaire are as follows.

1. Do you have any neurological or mental disorders?

2. Do you have a severe level of anxiety or depression?

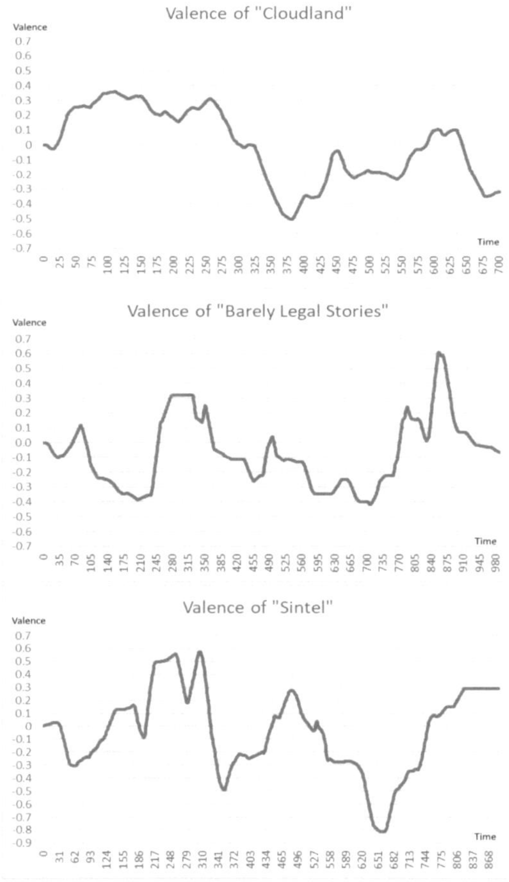


**Figure 5.**
*Valence plots of sample movies.*

**Figure 6.**
*Valence plots of negative movie (parafundit) and positive movie (tears of steel) we selected.*

3. Do you have hearing impairment that cannot be corrected?

4. Do you have any permanent disability or body injury that affects your walking posture?

5. Do you feel sick now? (e.g., fever, headache, stomachache)

6. If you have any problem with your health condition, please describe it.

According to this questionnaire, any subject who had a health issue could be excluded from participation. However, in this study, all subjects confirmed that they were healthy.

For the first dataset we proposed in [1], only health questionnaire listed above was used. In addition, for the second dataset proposed in this study, more questions were added to check for the dominant hand, dominant foot, and dominant brain side of each subject.

The dominant hand of each subject was determined using the modified version of *Flinders Handedness survey questions* published by *Left Handers Association of Japan* available online at https://lefthandedlife.net/faq003.html. All questions on this website were translated into Japanese and to make the questions more appropriate with Japanese culture. The questions of dominant hand questionnaire and their English translation are as follows.

1. 文字を書くとき、どちらの手でペン(筆記具)を持ちますか?

   When writing, which hand do you hold a pen (writing instrument)?

2. 食事をするとき、どちらの手でスプーンを持ちますか?

   When you eat, which hand do you hold the spoon?

3. 歯を磨くとき、どちらの手で歯ブラシを持ちますか?

   When brushing your teeth, which hand do you hold your toothbrush?

4. マッチを擦るとき、どちらの手でマッチ棒を持ちますか?

   When you rub a match, which hand do you hold the matchstick with?

5. 消しゴムで文字や図画を消すとき、どちらの手で消しゴムを持ちますか?

   When erasing letters and drawings with an eraser, which hand do you hold

6. お裁縫をするとき、どちらの手で縫い針を持ちますか?

   When sewing, which hand do you hold the sewing needle?

7. 食卓でパンにバターを塗るとき、どちらの手でナイフを持ちますか?

   When you put butter on bread at the table, which hand do you hold the knife?

8. 釘を打つとき、どちらの手で金づち(ハンマー)を持ちますか?

   When you hit a nail, which hand do you hold a hammer?

9. ジャガイモやりんごの皮をむくとき、どちらの手でピーラー (皮むき器) を持ちますか?

   When peeling potatoes or apples, which hand do you hold a peeler?

10. 絵を描くとき、どちらの手で絵筆やペンを持ちますか?

    When drawing, which hand do you hold a paintbrush or pen?

In each question, subjects can choose for left hand, right hand, and both hands. The score for each question is −1, +1, and 0 for left hand, right hand, and both hands, respectively. Total score for all questions was calculated for each subject to check the dominant hand of that subject. If the total score is −10 to −5, the subject is classified as left-handed. If the total score is −4 to +4, the subject is classified as both-handed, and if the total score is +5 to +10, that subject is right-handed.

Additionally, another questionnaire for checking the dominant foot for each subject was also used. Dominant foot was determined by using Chapman et al.'s Foot Dominant test questions, which are translated into Japanese language. The questions are available in Japanese at https://blog.goo.ne.jp/lefty-yasuo/e/37149f8d3105e9b43aa58c5925024915. The questions in Japanese and English translation are as follows.

1. サッカーボールを蹴る

   Which foot do you use to kick a soccer ball?

2. 缶を踏みつける

   Which foot do you use for trampling the can?

3. ゴルフボールを迷路に沿って転がす

   Which leg do you use to roll a golf ball along the maze?

4. 砂に足で文字を書く

   Which foot you use to write letters on the sand?

5. 砂地をならす

   Which foot do you use to smooth the sand?

6. 小石を足で並べる

   Which foot do you use to arrange the pebbles?

7. 足先に棒を立てる

   Which foot do you use to put a stick on your toes?

8. ゴルフボールを円に沿って転がす

   Which foot do you use to roll the golf ball along the circle?

9. 片足跳びをできるだけ速くする

   Which foot do you use to make one-legged jumps as fast as possible?

10. できるだけ高く足を蹴上げる

    Which foot do you use to kick your feet as high as you can?

11. 足先でこつこつリズムをとる

    Which foot do you use to take a rhythm with your feet?

The dominant foot was judged by checking the total score. If the subject's answer is left foot, the score is 3 points, right foot is 1 point, and both feet score is 2 points. In total, if the total score is 28 points or more, that subject was judged as left-footed. If the total score is less than 28 points, that subject was classified as right-footed.

Another questionnaire is for checking the dominant brain side. There are many dominant brain test questions available. In this study, we selected the arm and hand folding questions for testing the dominant brain side. There are two questions in this questionnaire. Subjects selected a picture that is matched with them for each question. The questions and pictures are from https://www.lettuceclub.net/news/article/194896/. Both questions are shown in Japanese and English as follows.

1. 自然に腕を組んでください。どのようになりましたか?

   Please fold your arm naturally. Which picture match with you?

2. 自然に手を組んでください。どのようになりましたか?

   Please fold your hand naturally. Which picture match with you?

Subjects were asked to select the picture of arm folding and hand folding that match with them. Hand folding test was used for testing the input brain, and arm folding test was used for testing the output brain. For hand folding, if the thumb of the right hand is below, the input brain is right side. If the thumb of the left and is below, the input brain is left side. For arm folding, if the right arm is below, the output brain is right side. If the left arm is below, the output brain is left side. The pictures for the subjects to select in the questionnaire are shown in **Figure 7**.

After finishing the health questionnaire, informed consent, dominant hand questionnaire, dominant foot questionnaire, and dominant brain side questionnaire, each subject was instructed to walk in a circular pattern inside the walking area marked by the black tape on the floor. Subjects are free to choose the direction they want to walk between clockwise or counter-clockwise. Also, subjects could switch the direction anytime when they want during each walking trial. The following are all walking trials each subject was asked to walk.

1. Walk in the rectangle walking area for 3 minutes as a rehearsal walk

2. Wear HoloLens 2 that showed nothing and walk in the walking area for 3 minutes as another rehearsal walk

3. Watch neutral video on HoloLens 2 while walking in the rectangle walking area

4. Watch the first emotional video (positive/negative video) on HoloLens 2 while walking in the rectangle walking area

5. Watch the second emotional video (negative/positive video) on HoloLens 2 while walking in the rectangle walking area

The intention of the first rehearsal walk is to make the subjects to be familiar with the room environment and the walking space. For the second rehearsal walk with HoloLens 2 showing nothing, as we found from [31–33], if the subjects have never

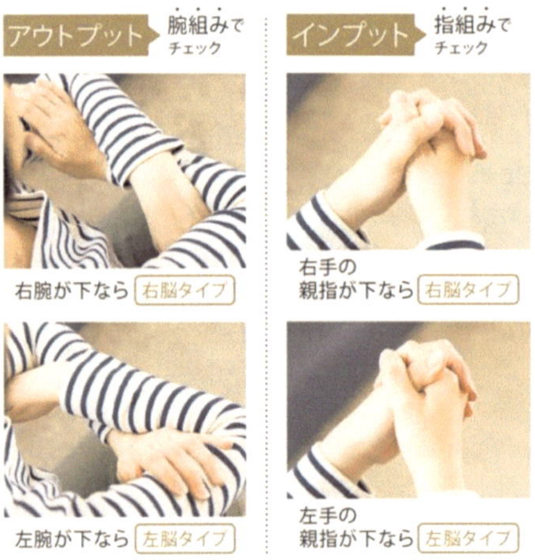


**Figure 7.**
*Arm and hand folding test questions (Source: https://www.lettuceclub.net/news/article/194896/).*

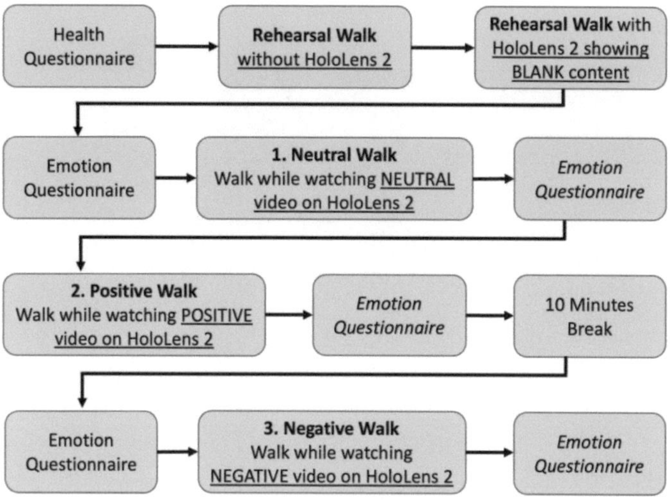


**Figure 8.**
*Data collection process.*

used smart glasses before, gait performance can be unstable. Therefore, we asked each subject to take another rehearsal walk to make the subject to feel familiar with walking and wearing HoloLens 2. After two rehearsal walks, we showed the neutral video on HoloLens 2 and ask each subject to start walking when the video starts and stop walking when the video stops. Then, we showed the first emotional video on HoloLens 2 and asked the subjects to walk in the same procedures as the first video. After this emotional video ended, we asked the subjects to go for a break for 10 minutes to reset their emotion to normal condition. Finally, we showed the second emotional video on HoloLens 2 and asked the subjects to walk while watching the last video. The first emotional video and the second emotional video were swapped between positive video then negative video, and negative video then positive video. Overall process for data collection of the first dataset is shown in **Figure 8**. For the second dataset, questionnaire for dominant hand, dominant foot, and dominant brain side was conducted after the health questionnaire and before the first rehearsal walk.

Furthermore, subjects were asked to report their perceived emotion after finishing neutral walk, positive walk, and negative walk. The questions are as follows.

- Please choose your current feeling: Happy, Sad, Neither (Not Sad & Not Happy)

- How intense of your feeling: 1 (Very Little) to 5 (Very Much)

In the first dataset, only self-reported emotion questionnaire was used after neutral walk, negative walk, and positive walk. In the second dataset, we added another question after the last self-reported questionnaire, that is, after finishing the last walking trial. As we are unsure that the subjects can walk naturally while they are watching videos on HoloLens 2 or not, we added a question asking them whether they can walk naturally while using HoloLens 2 and asked them to explain the reason.

Sample screenshots of a subject walking in circular pattern while watching a video on HoloLens 2 are shown in **Figure 9**. A sample image of a subject wearing HoloLens 2 and OptiTrack motion capturing suit with markers is shown in **Figure 10**.

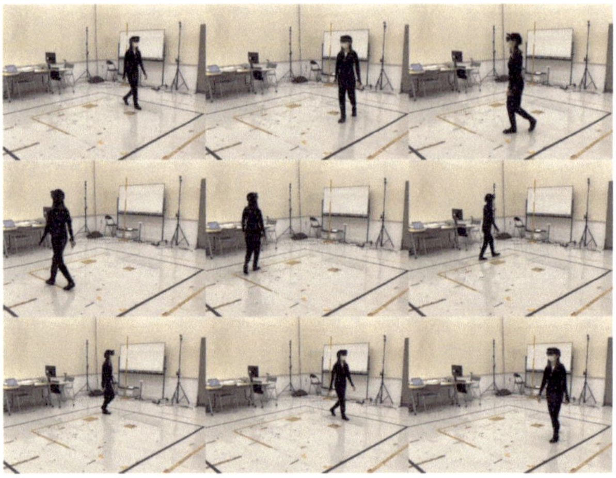

**Figure 9.**
*Samples of walking in the recording area.*

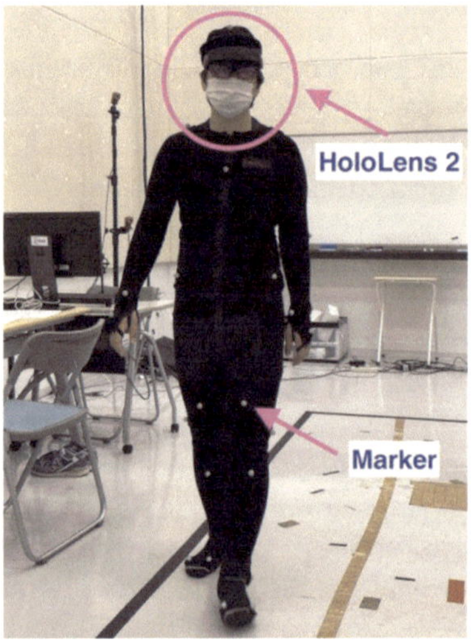


**Figure 10.**
*A Subject Wearing HoloLens 2 and OptiTrack Motion Capture Suit with 37 Markers.*

## 4. Result and discussion

Two emotional gait datasets were collected. The first dataset proposed in [1] contains 49 subjects including 41 men and 8 women. The average age is 19.69 years with 1.40 years standard deviation. The average height is 168.49 centimeters with 6.34 centimeters standard deviation. The average weight is 58.88 kilograms with 10.84 kilograms standard deviation. In total, there are 147 walking trials in this dataset. As the order of emotional videos shown to each subject was swapped, we have 24 subjects watched negative movie before positive movie (neutral – > negative – > positive), and 25 subjects watched positive movie before negative movie (neutral – > positive – > negative). For the emotion perceived by the subjects from the self-reported emotion questionnaire, there are 44 sad walking trials, 44 happy walking trials, and 59 neither walking trials. Comparison between expected emotion, which is the annotated emotions of the videos (negative, positive, neutral), and the reported emotion, which is the emotion reported by the subjects after finished walking (happy, sad, neither), is shown in **Table 2** and **Figure 11**.

As we also performed another data collection in addition to the first dataset, our second dataset contains 23 subjects including 10 men and 13 women. The average age is 19.91 years. The standard deviation of age is 3.04 years. The average height is 164.93 centimeters, and the standard deviation of height is 9.58 centimeters. The average weight is 57.32 kilograms, with 11.32 kilograms standard deviation. In total, this dataset consists of 69 walking trials. The order of emotional videos shown to each subject was also swapped as same as the first dataset. In this dataset, there are 12 subjects who watched negative movie before positive movie (neutral – > negative – > positive), and 11 subjects who watched positive movie before negative movie (neutral – > positive – > negative). Reported emotion that the subjects perceived compared with expected emotion is listed in **Table 3** and **Figure 12**. In this dataset, we also collected the dominant hand, dominant foot, and dominant brain side. The results of these questionnaire are as follows.

- **Dominant Hand:** 10 left-handed subjects, 8 right-handed subjects, and 5 both handed subjects

| Stimuli\Reported Emotion | Happy | Sad | Neither |
|---|---|---|---|
| Positive Movie | 12 | 23 | 14 |
| Negative Movie | 13 | 19 | 17 |
| Neutral Movie | 19 | 2 | 28 |

**Table 2.**
*Comparison of expected emotion and reported emotion (first dataset).*

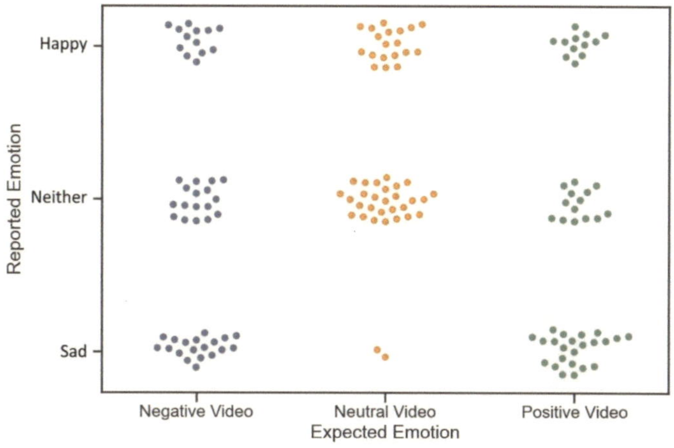


**Figure 11.**
*Plots of expected emotion and reported emotion (first dataset).*

| Stimuli\Reported Emotion | Happy | Sad | Neither |
|---|---|---|---|
| Positive Movie | 10 | 7 | 6 |
| Negative Movie | 0 | 19 | 4 |
| Neutral Movie | 12 | 3 | 8 |

**Table 3.**
*Comparison of expected emotion and reported emotion (second dataset).*

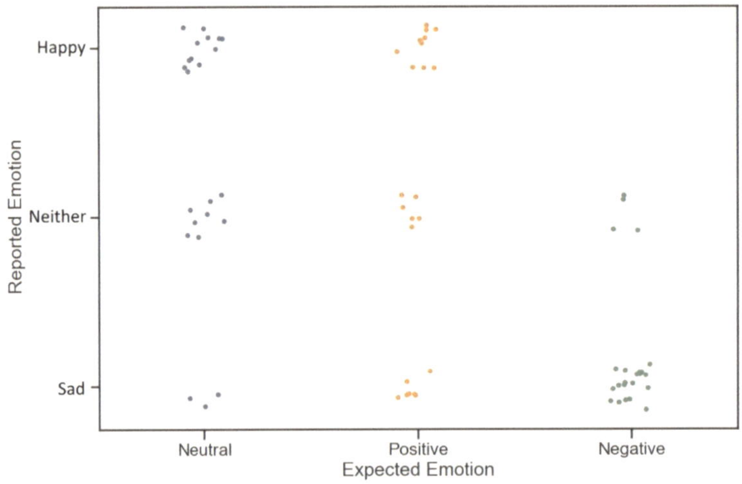


**Figure 12.**
*Plots of expected emotion and reported emotion (first dataset).*

- **Dominant Foot:** 8 left-footed subjects, 15 right-footed subjects
- **Dominant Brain Side:** 3 left-input/left-output subjects, 7 left-input/right-output subjects, 7 right-input/left-output subjects, and 6 right-input/right-output subjects

Dominant hand, dominant foot, and dominant brain side data will be useful in the future when this dataset is used for emotion recognition and analysis of body movements is performed.

According to **Table 2** and **Figure 11** that show the comparison of expected emotion and reported emotion for the first dataset, we found that not all subjects feel the emotions we want them to feel. That is, positive video cannot make everyone feel happy, and negative video cannot make everyone feel sad. From positive video, the number of subjects who feel sad is almost twice from the subjects who feel happy; that is, 12 subjects feel happy while 23 subjects feel sad. For negative video, more subjects feel sad than happy; that is, 19 subjects feel sad while 13 subjects feel happy. For neutral video, the results are quite random since we intended to make it not inducing any emotion. Therefore, most subjects feel neither for neutral video. Other reported emotions of neutral video including happy and sad can occur because of other causes. For example, if the subjects have never used HoloLens 2 before, walking while watching a video on HoloLens 2 can make them feel happy, sometimes, if the subjects feel uncomfortable while walking and watching a video on HoloLens 2 at the same time, it is possible that they will feel sad after watching neutral video.

For the second dataset, **Table 3** and **Figure 12** show that the reported emotions for positive movie are 10 subjects feel happy and 7 subjects feel sad, which are not so different. These results reveal that the emotion induction for positive movie is not so effective. In the first dataset, a lot more subjects feel sad after watching positive movie, which is opposite to the emotion we want them to feel. In the second dataset, more subjects feel happy than feel sad after watching positive movie. But the numbers are not much different. This still means that emotion induction for positive video as stimuli is not effective enough even though the number of happy subjects is higher than sad subjects unlike the first dataset. Next, when we consider the negative movie, in the first dataset, 13 subjects feel happy and 19 subjects feel sad. In comparison with the second dataset, no one feels happy but 19 subjects feel sad. The results show that emotion induction for negative video in the second dataset is more effective than the first dataset although the stimulus we used for negative video is the same movie. One possible reason is that the subjects in the second dataset are more sensitive to negative movie than the subjects in the first dataset. Moreover, neutral video for both dataset results in random reported emotion for both datasets. This is the good outcome showing that the neutral video did not induce any emotion as we expect this video to be.

From all results of comparison between expected and reported emotions, we can see that the reported emotions are not similar to the expected emotions, which are the annotated emotions of the video stimuli. There are several possible causes; for instance, some subjects might be more sensitive to feel sad when they saw some stories. That is, some subjects can feel very sad, whereas other subjects can feel little sad, not sad and not happy (neutral), or feel happy when they saw the negative movie. This phenomenon is normal since different people have different emotion perception. This explanation is also valid for positive movie. Although the annotated emotion of this movie is positive, some subjects can feel happy while other subjects feel sad. Another possible reason is that, sometimes, the subjects cannot fully understand the contents of the movies because they watched the movies and walked at the same time. Therefore, subjects need to concentrate on walking in addition to watching movies. This makes some subjects cannot completely understand the content of the movies, so the reported emotion is opposite from the desired emotion we want them to feel. Other possible explanation is that some components or stories in the positive movie can make some subjects feel sad. For example, some music or video scenes might be very intense and some subjects are sensitive to these contents. This explanation is still related to the previous one. Additionally, individual preferences are also importance issues that should be considered. For example, if some subjects do not like sci-fi movie that we used as the positive stimuli, that movie can make them feel sad because this movie is the kind they do not like. Music soundtrack of the movies is also the reason that makes the perceived emotions different from the emotion we expected. If the subjects like the music, they can feel happy even the movie is negative movie, and if the subjects do not like the music, they can feel sad even the movie is positive movie. Lastly, if the subjects did not feel well when they watch the movie during walking, for example, some subjects feel motion sickness, or some subjects feel bored, the perceived emotions will be inaccurate and different from the emotions we expected. Because of this reason, we asked the subjects whether they can walk naturally or not after they finished walking.

In the first dataset, we did not have this information. But for the second dataset, seven subjects answered that they can walk naturally, eight subjects answered that they cannot walk naturally, and eight subjects answered they are unsure. If we consider their explanation, the subjects who answered they can walk naturally have positive feedbacks. The followings are some examples.

- 映像に集中していたから

  Because I was concentrating on the video

- 歩きにくさを感じなかったから

  Because I did not find it difficult to walk

- 飽きなかったから!

  I did not get tired of it!

- 視界が完全に隠されてたわけではなかったため。あしもとは見えてたので比較的歩きやすかった。

- Because the view wasn't completely hidden. I could see the foot, so it was relatively easy to walk.

Unfortunately, the subjects who answered that cannot walk naturally have negative feedbacks with walking while watching videos on HoloLens 2. These are some examples.

- 音や映像に気を取られたから

  Because I was distracted by the sound and image

- 途中でフラついたり、まっすぐ歩けなかったりしたからです。

  Because I was fluttering on the way and I could not walk straight

- 映像に集中していて、たまに枠線を超えそうになったから。

  Because I was concentrating on the video and sometimes I almost crossed the border.

- 歩く範囲が小さいため

  Because the walking range is small

Even some subjects answered that they feel unsure, some feedbacks are also negative.

- 時々眠たかったから

  Sometimes I wanted to sleep

- 映像を見ながら歩くのが少し難しかった

  It was a little difficult to walk while watching the video

- 枠外には出なかったが、円状を単調に歩いていたので、目がくらんで、不自然に歩いていたかもしれないから。

I did not go out of the frame, but I was walking monotonously in a circle, so I might have been dazzled and walked unnaturally.

There are some positive feedbacks from unsure subjects. These answers show that they are walk unconsciously so we can collect the very natural walking styles.

- 自分がどう歩いていたか気にしていなかった

  I did not care how I was walking

- 何も考えていなかったため、自然だったかはわからないです

  I did not think about anything, so I do not know if it was natural

From these answers, we can see that some subjects have difficulties with watching videos on HoloLens 2 while walking at the same time. These are very reasonable explanations why emotion induction is not effective, and the reported emotions are much different from expected emotions. In the first dataset, we did not collect these data, hence, we cannot know that the subjects can walk naturally or not, and we cannot know how they feel after watching videos on HoloLens 2 while walking, but for the second dataset, we have these data, and they are very useful.

By using Microsoft HoloLens 2 for displaying emotional videos, there are several things we found and worth for consideration.

- This kind of emotion induction method can be used to simulate real-time emotion. However, using entire movies is not so effective.
    - The subjects need to pay too much attention with the movie, so some subjects cannot walk naturally, some subjects have motion sickness, and some subjects feel bored because the movies are too long.
    - If subjects decide to give priority to walking more than watching movies, some subjects cannot fully understand the movie content.
- HoloLens 2 can be used for showing contents to the subjects to simulate real-time emotion, but showing entire movie is not a good idea because of many negative feedbacks from the subjects.
    - Showing short movies instead of full movie or using some animation as mixed reality agents (VR/AR) should be better ideas to induce real-time emotion of the subjects.
- Asking subjects to report their perceived emotion using the self-reported questionnaire is very important since the desired emotion we expected can be different from the emotion they reported.
- Displaying video during walking can induce natural emotion because several subjects are unconscious that their walking postures change or not.
- Currently, trust the subjects what they feel from the self-reported questionnaire is better than using the emotion tag from the stimuli.

## 5. Conclusion

To summarize, this study extends our previously proposed emotion induction and data collection method [34]. In conventional emotion induction, emotional videos are shown to the subjects before walking using a computer screen or television. In our method, emotional videos are shown on Microsoft HoloLens 2, which is the latest smart glasses technology. We found that displaying emotional videos on HoloLens 2 while walking can make the subjects express emotions on their gaits unconsciously while walking. Subjects can also see the room environment and the stimuli contents on HoloLens 2 at the same time. Some subjects think it is easy to walk while watching videos on HoloLens 2, while some subjects say that it is difficult to focus on waking and paying attention with video contents on HoloLens 2 at the same time. Our goal of this study is to simulate a real-time emotion while walking, however using full-length movies may not be a good idea because there are some negative feedbacks from our participants. For the walking direction, using a non-straight walking path will make the data collection become more realistic since capturing human gaits in reality is difficult to capture only straight and clean walking data. Therefore, if the emotion recognition system was developed and tested using non-straight walking gait data, opportunities to deploy this system in real-world scenario are increased. Additionally, expected emotions, which are the annotated emotion of the stimuli, should not be used to tag a walking trial since the stimuli emotions can be opposite to the emotions that the subjects perceived. Asking the subjects to report their actual feelings after walking is the best way we can do for now. In this study, OptiTrack motion capturing system was used for capturing gait data, but it is not mandatory to use marker-based systems such as OptiTrack or Vicon. Using marker-less motion capturing device, for example, Microsoft Kinect is also possible. Even a standard video camera or mobile phone camera with pose-estimation software such as OpenPose can also be used to capture body movement data for emotion recognition by gait analysis. In summary, this study investigates the possibility of performing emotion recognition and analysis by using smart glasses to induce emotions of subjects. The results show that emotion recognition from human gaits can be performed and is very useful in many circumstances. Since emotion recognition is an example of tangible applications of intelligence video surveillance, methods for inducing human emotions should be also considered. To develop an effective emotion recognition system as a part of intelligence video surveillance, obtaining a high-quality dataset is an important factor that should be focused on.

## 6. Acknowledgment

The author would like to thank all participants who joined our both experiments. Additionally, we appreciated the help and support from all members of Kaoru Sumi Laboratory at Future University Hakodate, who supported and assisted in both experiments such as experiment venue setup, equipment setup, experimental design, translation of all documents, and interpretation between Japanese and English during the entire experiments processes.

## Funding statement

This work was supported by JST Moonshot R&D Grant Number JPMJMS2011.

## Author details

Nitchan Jianwattanapaisarn[1,2], Kaoru Sumi[1]* and Akira Utsumi[2]

1 School of Systems Information Science, Future University Hakodate, Hokkaido, Japan

2 Interaction Science Laboratories, Advanced Telecommunications Research Institute International, Kyoto, Japan

*Address all correspondence to: kaoru.sumi@acm.org

## References

[1] Picard RW. Affective Computing. MIT press; 2000

[2] Tiam-Lee TJ and Sumi K. Analysis and prediction of student emotions while doing programming exercises. In: International Conference on Intelligent Tutoring Systems. Springer. 2019. pp. 24–33

[3] Bouchrika I. A survey of using biometrics for smart visual surveillance: Gait recognition. In Surveillance in Action. Cham: Springer; 2018. pp. 3-23. DOI: 10.1007/978-3-319-68533-5_1

[4] Anderez DO, Kanjo E, Amnwar A, Johnson S, Lucy D. The rise of technology in crime prevention: Opportunities, challenges and practitioners perspectives. 2021. arXiv preprint arXiv:2102.04204

[5] Montepare JM, Goldstein SB, Clausen A. The identification of emotions from gait information. Journal of Nonverbal Behavior. 1987;**11**(1):33-42

[6] Khamsemanan N, Nattee C, Jianwattanapaisarn N. Human identification from freestyle walks using posture-based gait feature. IEEE Transactions on Information Forensics and Security. 2017;**13**(1):119-128

[7] Limcharoen P, Khamsemanan N, Nattee C. View-independent gait recognition using joint replacement coordinates (jrcs) and convolutional neural network. IEEE Transactions on Information Forensics and Security. 2020;**15**:3430-3442

[8] Limcharoen P, Khamsemanan N, Nattee C. Gait recognition and re-identification based on regional lstm for 2-second walks. IEEE Access. 2021;**9**: 112057-112068

[9] Kitchat K, Khamsemanan N, Nattee C. Gender classification from gait silhouette using observation angle-based geis. In: 2019 IEEE International Conference on Cybernetics and Intelligent Systems (CIS) and IEEE Conference on Robotics, Automation and Mechatronics (RAM). IEEE. 2019. pp. 485–490

[10] Isaac ER, Elias S, Rajagopalan S, Easwarakumar K. Multiview gait-based gender classification through pose-based voting. Pattern Recognition Letters. 2019;**126**:41-50

[11] Janssen D, Schöllhorn WI, Lubienetzki J, Fölling K, Kokenge H, Davids K. Recognition of emotions in gait patterns by means of artificial neural nets. Journal of Nonverbal Behavior. 2008;**32**(2):79-92

[12] Roether CL, Omlor L, Christensen A, Giese MA. Critical features for the perception of emotion from gait. Journal of Vision. 2009;**9**(6):15-15

[13] Karg M, Kühnlenz K, Buss M. Recognition of affect based on gait patterns. IEEE Transactions on Systems, Man, and Cybernetics, Part B (Cybernetics). 2010;**40**(4):1050-1061

[14] Barliya A, Omlor L, Giese MA, Berthoz A, Flash T. Expression of emotion in the kinematics of locomotion. Experimental Brain Research. 2013;**225** (2):159-176

[15] Venture G, Kadone H, Zhang T, Grèzes J, Berthoz A, Hicheur H. Recognizing emotions conveyed by human gait. International Journal of Social Robotics. 2014;**6**(4):621-632

[16] Li B, Zhu C, Li S, Zhu T. Identifying emotions from non-contact gaits information based on microsoft kinects.

IEEE Transactions on Affective Computing. 2016;**9**(4):585-591

[17] Li S, Cui L, Zhu C, Li B, Zhao N, Zhu T. Emotion recognition using kinect motion capture data of human gaits. PeerJ. 2016;**4**:e2364

[18] Zhang Z, Song Y, Cui L, Liu X, Zhu T. Emotion recognition based on customized smart bracelet with built-in accelerometer. PeerJ. 2016;**4**:e2258

[19] Chiu M, Shu J, Hui P. Emotion recognition through gait on mobile devices. In: 2018 IEEE International Conference on Pervasive Computing and Communications Workshops (PerCom Workshops). IEEE. 2018. pp. 800–805

[20] Quiroz JC, Geangu E, Yong MH. Emotion recognition using smart watch sensor data: Mixed-design study. JMIR Mental Health. 2018;**5**(3):e10153

[21] Xu S, Fang J, Hu X, Ngai E, Guo Y, Leung V, et al. Emotion recognition from gait analyses: Current research and future directions. arXiv preprint arXiv: 2003.11461. 2020.

[22] Lemke MR, Wendorff T, Mieth B, Buhl K, Linnemann M. Spatiotemporal gait patterns during over ground locomotion in major depression compared with healthy controls. Journal of Psychiatric Research. 2000;**34**(4–5):277-283

[23] Michalak J, Troje NF, Fischer J, Vollmar P, Heidenreich T, Schulte D. Embodiment of sadness and depression—gait patterns associated with dysphoric mood. Psychosomatic Medicine. 2009; **71**(5):580-587

[24] Gross MM, Crane EA, Fredrickson BL. Effort-shape and kinematic assessment of bodily expression of emotion during gait. Human Movement Science. 2012;**31**(1):202-221

[25] Destephe M, Maruyama T, Zecca M, Hashimoto K, Takanishi A. The influences of emotional intensity for happiness and sadness on walking. In: 2013 35th Annual International Conference of the IEEE Engineering in Medicine and Biology Society (EMBC). IEEE. 2013. pp. 7452-7455

[26] Sun B, Zhang Z, Liu X, Hu B, Zhu T. Self-esteem recognition based on gait pattern using kinect. Gait & Posture. 2017;**58**:428-432

[27] Halovic S, Kroos C. Not all is noticed: Kinematic cues of emotion-specific gait. Human Movement Science. 2018;**57**: 478-488

[28] Sadeghi H, Allard P, Duhaime M. Functional gait asymmetry in able-bodied subjects. Human Movement Science. 1997;**16**(2-3):243-258

[29] Kang GE, Gross MM. Emotional influences on sit-to-walk in healthy young adults. Human Movement Science. 2015;**40**:341-351

[30] Kang GE, Gross MM. The effect of emotion on movement smoothness during gait in healthy young adults. Journal of Biomechanics. 2016;**49**(16): 4022-4027

[31] Kim S, Nussbaum MA, Ulman S. Impacts of using a head-worn display on gait performance during level walking and obstacle crossing. Journal of Electromyography and Kinesiology. 2018;**39**:142-148

[32] Sedighi A, Ulman SM, Nussbaum MA. Information presentation through a head-worn display ("smart glasses") has a smaller influence on the temporal structure of gait variability during dual-task gait compared to handheld displays (paper-based system and smartphone). PLoS One. 2018;**13**(4):e0195106

[33] Sedighi A, Rashedi E, Nussbaum MA. A head-worn display ("smart glasses") has adverse impacts on the dynamics of lateral position control during gait. Gait & Posture. 2020;**81**: 126-130

[34] Jianwattanapaisarn N, Sumi K. Investigation of real-time emotional data collection of human gaits using smart glasses. Journal of Robotics, Networking and Artificial Life. 2022;**9**(2):159-170. DOI: 10.57417/jrnal.9.2_159

[35] Baveye Y, Dellandréa E, Chamaret C, Chen L. Deep learning vs. kernel methods: Performance for emotion prediction in videos. In: 2015 International Conference on Affective Computing and Intelligent Interaction (acii). IEEE. 2015. pp. 77–83

Chapter 9

# Combining Supervisory Control and Data Acquisition (SCADA) with Artificial Intelligence (AI) as a Video Management System

*Muhammad H. El-Saba*


## Abstract

The latest Video management systems (VMS) software relies on CCTV surveillance systems that can monitor a larger number of cameras and sites more efficiently. In this paper, we study the utilization of SCADA to control a network of surveillance IP cameras. Therefore, the video data are acquired from IP cameras, stored and processed, and then transmitted and remotely controlled via SCADA. Such SCADA application will be very useful in VMS in general and in large integrated security networks in particular. In fact, modern VMS are progressively doped with artificial intelligence (AI) and machine learning (ML) algorithms, to improve their performance and detestability in a wide range of control and security applications. In this chapter, we have discussed the utilization of existing SCADA cores, to implement highly efficient VMS systems, with minimum development time. We have shown that such SCADA-based VMS programs can easily incubate AI and deep ML algorithms. We have also shown that the harmonic utilization of neural networks algorithms (NNA) in the software core will lead to an unprecedented performance in terms of motion detection speed and other smart analytics as well as system availability.


**Keywords:** SCADA, distributed control systems (DCS), security, artificial intelligence (AI), face recognition, machine learning, video surveillance video analytics

## 1. Introduction

The Supervisory Control and Data Acquisition (SCADA) is a software overlay application, which is used on top of intelligent control networks. The control nodes are traditionally smart sensors and programmable logic controllers (PLCs) [1–3]. In fact, the SCADA industry started due to the need for a user-friendly front-end to control systems containing smart devices, devices, and PLCs. However, SCADA systems evolved rapidly and are now penetrating the reliable operation of modern infrastructures [2, 3] of smart cities. SCADA systems have made substantial progress over the recent years to increase their functionality, scalability, performance, and openness. As shown in **Figure 1**, the main components of a SCADA system are as follows:

- Multiple Remote Terminal Units (RTUs) or Smart sensors or PLCs.
- Master Station and Human machube interface (HMI) Computer(s).
- Communication infrastructure

Indeed, it is possible to purchase a SCADA system from a single supplier or tailor a SCADA system from different manufacturers, such as Siemens and Allen-Bradley PLCs. This chapter presents a method based on neural networks (NN) for monitoring and operating video surveillance systems (VMS), like those in traffic control networks and electronic plaza sites. The method suggests that the thresholds used for generating alarms can be adapted to each surveillance device (e.g., IP Camera). The intelligent SCADA method was actually utilized in other application fields, for example, in electrical power control and renewable energy systems [3].

In this chapter, we show how to exploit such existing SCADA programs to implement a wide-area video management systems (VMS), which incorporate state-of-the-art AI technologies, such as access control, intrusion detection, face recognition, license plate recognition, crowd detection, and city surveillance. These technologies have been implemented in our emerging VMS, Xanado [4], which is expected to have a unique value in identifying criminals and terrorists, patrolling highways, and in aiding forensics.

Such SCADA solutions are multi-tasking and are based upon a real-time database that is located on dedicated servers of the system. Such SCADA servers are responsible for data acquisition and handling (e.g., data polling, alarm checking, logging verification, and data archiving) on the basis of a set of chosen parameters.

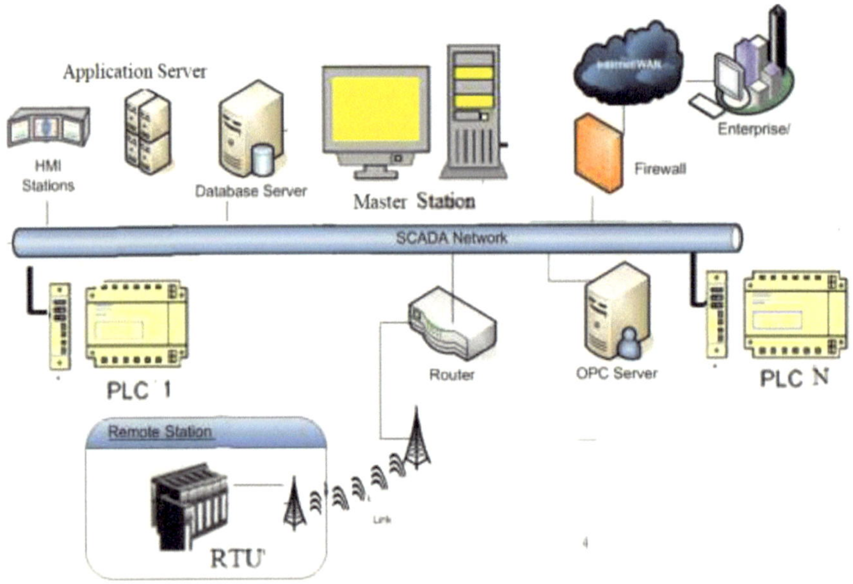


**Figure 1.**
*Conventional Architecture of a SCADA system, The Master Station refers to the servers and software responsible for communicating with Remote Terminal Units (RTUs), such as PLCs.*

*DOI: http://dx.doi.org/10.5772/intechopen.104766*

## 2. SCADA-based video surveillance solutions

One of the main advantages of SCADA systems is that it allows operators to visualize, in real time, what is happening in any particular industrial process, react to alarms, control processes, change configurations, and track information in real time. However, SCADA systems differ from distributed control systems (DCSs) that are generally found in industrial plant sites. While a DCS covers a plant site, a SCADA system covers much larger areas. Similarly, wide-area video monitoring requires large-scale monitoring systems, like those of SCADA networks. For operations that span several sites, it is important to have a central monitoring station that acts as eyes and ears across all sites. Central monitoring stations use diverse types of cameras and sets of technology to monitor and protect people and property, especially when personnel cannot be on site. It is important that these technologies work together to create a holistic monitoring system. Fortunately, SCADA architecture supports TCP/IP (Internet Protocol), UDP, or other IP-based communications protocols, which makes it ideal for video surveillance control with a network of IP cameras. In fact, SCADA systems have traditionally used combinations of direct serial buses or Ethernet or Wi-Fi connections to meet communication requirements, as well as IP over SONET (Synchronous Optical Network) at large sites. **Figure 2** depicts the architecture of SCADA-based VMS programs, which are employing neural network algorithms (NNA). The employed NNA aims to improve the control of the system by using an iterative supervised process. The objective is to determine and optimize the SCADA-VMS control parameters, for specific sites with specific surveillance devices. The chosen parameters, such as the favorite angles of PTZ cameras, the detection speed (of motion anomalies), and false arm causes, will help to increase the surveillance performance and system availability. In addition, the optimized system will minimize false alarms in a continuous adaptive manner, according to each site-specific equipment. Actually, the NNA is based on finding differences in the behavior of the surveillance system over time. The iterative process starts with a database of the stored SCADA-VMS database, as shown in **Figure 2**.

The SCADA-based VMS programs, with NNA, can help in this context and can easily incorporate the following features and intelligent analytics:

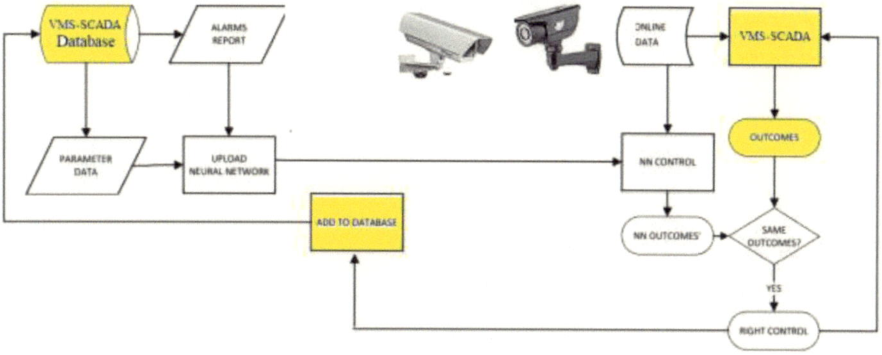


**Figure 2.**
*Block diagram of SCADA-based VMS, with neural network algorithms (NNA).*

1. Object detection and tracking in surveillance

2. Face recognition systems for access control

3. Automatic Number Plate Recognition (ANPR)

4. Incident management to identify, analyze and correct dangerous situations.

### 2.1 Object detection and tracking

The process of identifying objects in an image and finding their position is known as object detection. **Figure 3** depicts the object detection and identification tasks. This activity benefitted a lot from the field of computer vision assisted by AI. As shown, the trained model using deep learning must be evaluated for its performance on some data called as test dataset' [5–9].

### 2.2 Face recognition

Face identifiers offer advantages in access control, safety, security, retail stores, and traffic control. Facie recognition is actually an analytic program that identifies persons from their facial features in an image or video surveillance.

Face recognition programs are usually utilizing AI to quickly identify and interpret complex figures. When comparing a face image to a database of previously stored images of known faces, the AI algorithm can then determine the best match. In fact, the face recognition and analysis algorithms have enabled security systems to capture many wanted criminals and stop many crimes.

As facial recognition increases in efficacy, the number of other applications will also increase (e.g., in banking, retail stores, and means of transportation). Thanks to deep learning-based AI algorithms, face analysis is not only able to identify with high accuracy, but also provides extraordinary analytic capabilities. For instance, it can now detect the criminal behavior, from his/her mood and gests. Thanks to 3D/4D digital signal processing (DSP), with AI, face recognition technology will be expanding in identifying terrorists and criminals' actions, by incorporating motion detection algorithms.

### 2.3 Automatic number plate recognition (ANPR)

The ANPR is analytic software that reads vehicle plates and automatically matches them to recognized vehicle license plates, without the need for human

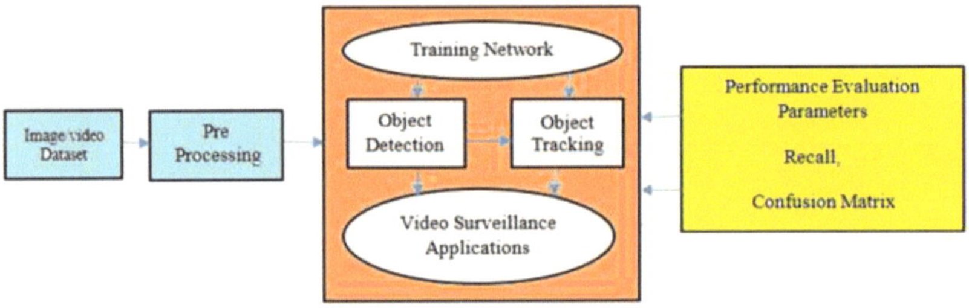


**Figure 3.**
*Block diagram of object detection and identification y artificial intelligence.*

intervention. Therefore, ANPR offers accurate identification and safety of vehicle access and traffic control. ANBR is usually implemented using optical character recognition (OCR) and convolutional neural networks (CNN). Actually, CNN is a widely used neural network architecture for computer vision tasks. The CNN automatically extracts important features on images. More details about CNN are provided below in Section 4.

## 3. Artificial intelligence in video surveillance and video analytics

Artificial intelligence (AI) is nowadays revolutionizing video management systems (VMS) and the ways of securing smart premises and smart cities by video surveillance and control. In fact, AI-based technology can promote security and surveillance equipment by enhancing object detection and motion interpretation as well as providing analytics with increasingly reliable data. For instance, the reduction of false alarms in security systems is one of the major benefits of AI-based tools and algorithms. For instance, AI-CCTV cameras are networked IP cameras that deliver advanced analytical functions, such as face recognition, vehicle classification, car counting, license plate recognition (LPR), and other traffic analytics. Advanced video analytics software is built into the camera and recorder, which then enables artificial intelligence functions. Some AI algorithms are rule-based and others are self-learning. Like typical CCTV cameras, AI-CCTV stores information so any incidents can be reviewed. However, AI CCTV can detect and send alerts in real time. In particular, SCADA legacy can help a lot in large-sites and wide-area VMS systems.

The artificial intelligence (AI) tools are heavily dependent on neural networks (NN) and computer vision [5]. As shown in **Figure 4**, a neural network (NN) is a system of software or hardware, which mimics the operation of human brain neurons. Therefore, an NN is simply a group of interconnected layers of a perceptron. Note that an NN has multiple hidden layers, and each layer has multiple nodes. The neural network takes the training data in the input layer and forwards it through hidden layers, on the basis of specific weights at each node [10]. Therefore, it returns an output value to the output layer. The inputs to nodes in a single layer have adaptable weights that affect the final output prediction.

There are a lot of different kinds of neural networks that are used in machine learning projects. There are recurrent neural networks, feed-forward neural networks, and convolutional neural networks (CNNs). It can take some time to properly tune a neural network to get consistent, reliable results. Testing and training your NN is important before deciding which parameters (input features of a face image) are important in your recognition model.

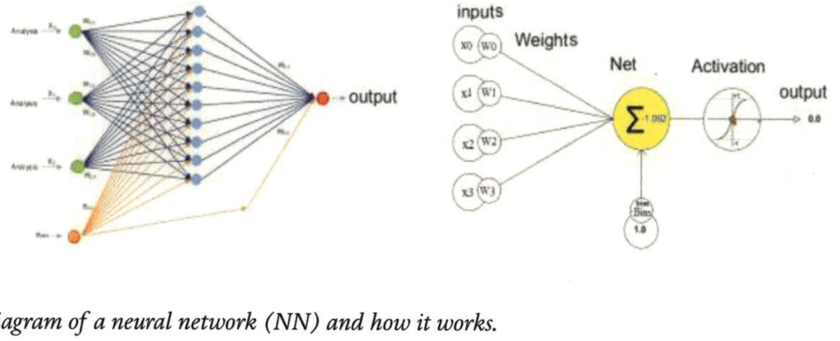


**Figure 4.**
*Schematic diagram of a neural network (NN) and how it works.*

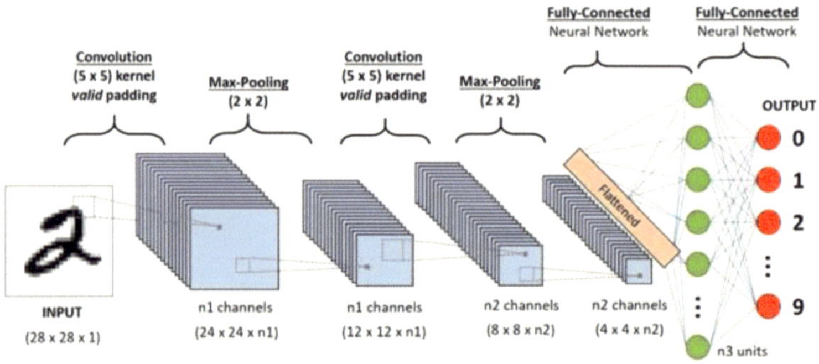


**Figure 5.**
*Schematic diagram of the layers of a convolutional neural network (CNN) showing its classification sequence of a handwritten character (in ASCII).*

## 4. Deep learning-based video surveillance solutions

In deep learning, a convolutional neural network (CNN) is a sort of NNs; which is particularly useful in video surveillance projects. The advantage of deep learning-(DL)-based algorithms with respect to legacy computer vision algorithms is that DL systems can be continuously trained and improved with updated datasets.

In DL, a convolutional neural network (CNN) is a type of NN, commonly used in image recognition and processing, with emphasis on machine vision of images and video. As shown in **Figure 5**, the layers of a CNN consist of an input layer, an output layer, and a hidden layer that includes multiple convolutional layers, pooling layers, fully connected layers, and normalization layers.

Deep learning systems have shown a remarkable ability to detect undefined or unexpected events. This feature has the true potential of significantly reducing false alarm events that happen in many security video analytics systems. Many applications have shown that deep learning systems can "learn" to achieve 99.9% accuracy for certain tasks, in contrast to rigid computer algorithms where it is very difficult to improve a system past 95% accuracy [4].

## 5. Xanado program

Xanado is a video management system (VMS) software, designed for large-scale and high-security installations. It is built as a client-server DCS to ensure end-to-end protection of video integrity and boost the overall performance of existing hardware. In addition to central management of all data servers, IP cameras, and users in a multi-site setup, Xanado includes an integrated video wall size for operators demanding overall awareness of any event. The software supports failover recording servers making it the perfect choice for mission-critical installations that require continued access to video recording in case of a server failure. Xanado is ideal for installations with 24/7 operation requirements and can run on high-speed recording engines (NVR), making it suitable for monitoring airports, banks, traffic control, as well as smart city surveillance.

The general system architecture is shown in **Figure 6**. As shown in Figure, the management data server lies in the center of the VMS. It holds the main application and handles the system configuration. Note that the recording server is responsible

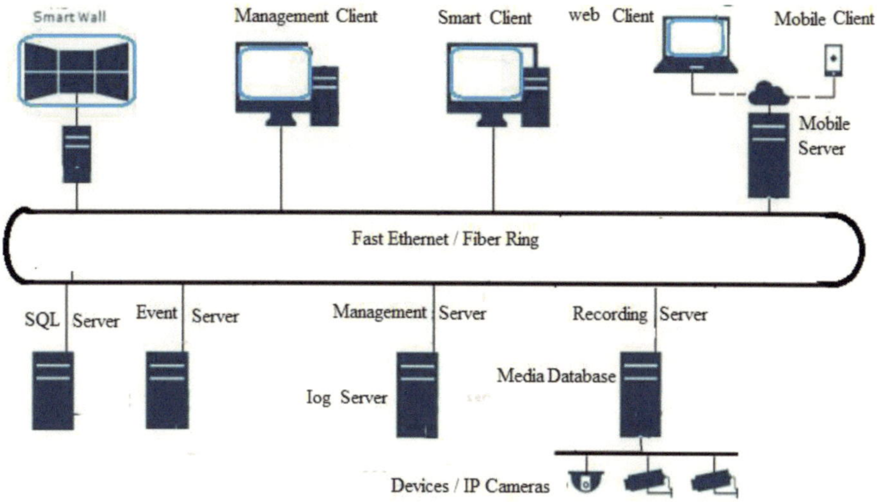


**Figure 6.**
*Xanado general system architecture.*

for all communication, recording, and event handling related to devices, such as cameras and I/O modules. The system stores the video in a customized database. The management server, event server, and log server use an SQL server to store configuration, alarms, and log events. As the VMS is designed for a large-scale operation, the Management Client may run locally or remotely, for centralized administration.

The smart client nodes (SCN) are working as follows. SCN connects to the management server and attempts to log in. The management server tries to authenticate the user and the user-specific configuration is retrieved from the SQL database. Therefore, the login is granted and the configuration is sent to the SCN. Live video streams are then retrieved from cameras by the recording server. The recording server sends a multicast stream to the multicast-enabled network. This requires that all switches handling the data traffic between the SCN and the recording server must be configured for multicast.

We adopt the ONVIF standard [11] for full video interoperability in multi-vendor installations to ensure information exchange by a common protocol. The ONVIF protocol profiles are collections of specifications for interoperability between ONVIF compliant devices, such as cameras and NVR.

Of course, Xanado VMS has many add-on modules and can be tailored to several specific applications. In the following subsection, we describe an application of our VMS to the case of Electronic Toll Plaza control, which is utilized nationwide in highway traffic control.

## 6. Case study: toll plaza control

Toll plazas are utilized on traffic highways to collect fees from passing vehicles. The toll plaza consists of six zones—an approach zone, queue area, toll lanes, the toll island, departure zone, a bailout lane, and, in some cases, a terminal supervisor. The so-called Electronic Toll Collection (ETC) enables toll collection without delay or total stop of vehicles. This section deals with the equipment and software to be

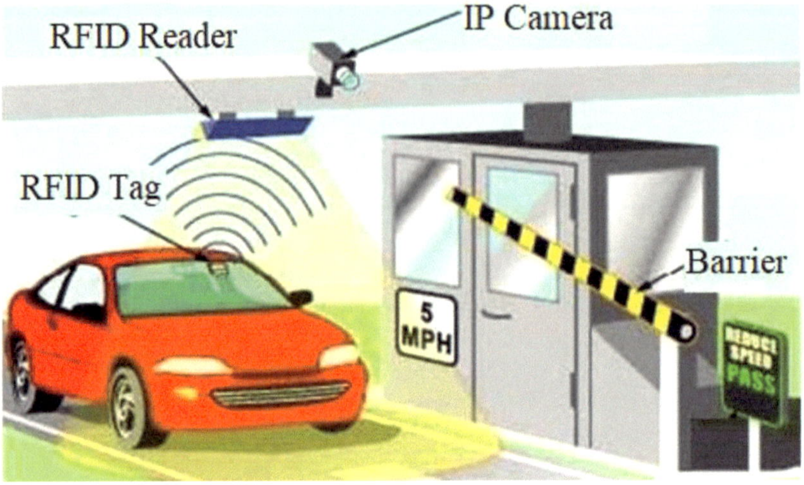


**Figure 7.**
*Schematic of a booth in an electronic toll office, which communicates with a passing vehicle.*

installed on the roadside of toll plaza networks. We shortly depict the basic concepts for installing the system of electronic toll collection. **Figure 7** is a schematic of a single booth and a lane of an electronic toll plaza.

### 6.1 Main specifications of a smart toll-plaza control system

A smart toll collection system has the following four components. The first three components are usually installed at the toll booths. The later backend is installed in the control room and is usually connected with manages the complete toll collection process [12].

- Vehicle Tracking System: This is a set of surveillance cameras, vehicle identification, and hardware sensors (e.g., RFID readers and loop detectors).

- Vehicle Enforcement Devices, such as barriers.

- Billing Terminals (POS), for manual fees collection of passing vehicles.

- Toll Management System: This system is responsible for processing the authentication and billing data of passing vehicles and dispatching this data to the headquarter office.

### 6.2 Toll plaza components equipment

The toll system comprises of Lane System and Plaza System, integrated into an architecture that facilitates easy and accurate toll collection. The following **Figures 8** and **9** depict the lane and plaza equipment. The OHLS (Over Head Lane Signal) is often required. The so-called AVCC (auto vehicle classification system ) is needed to determine the different fares of different vehicles if lanes have no signage with the type of passing vehicle. AVCC Systems may be Treadle systems that use a combination of vehicle magnetic loop, height sensors, and piezo sensors. Alternatively, AVCC may be IR-based.

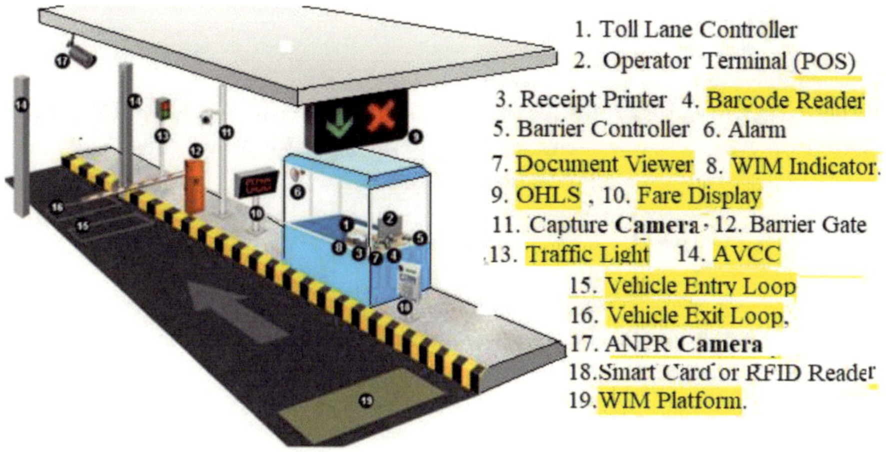


**Figure 8.**
*Schematic of toll-plaza lane and booth equipment.*

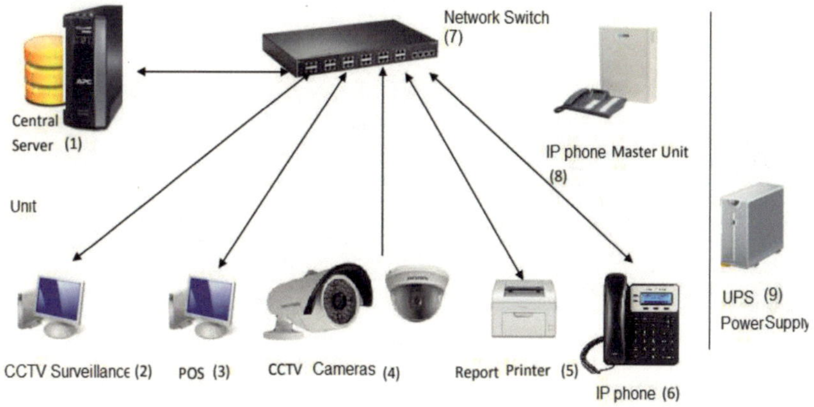


**Figure 9.**
*Schematic of plaza equipment.*

Also, the PBX telephone, which may be required inside each booth, is not shown. The WIM (Weigh-In-Motion) platform, which senses and records the vehicle weight, is dedicated to truck traffic control. Additional cameras should be installed inside each booth. In fact, a major challenge faced by any concessionaire in operating toll roads is the prevention of revenue leakage that goes very high like 20% of the daily collection in the situation of un-monitored systems.

### *6.2.1 Booth & Lane equipment list*

a-Booth Equipment: 1. Toll Lane Controller; 2. Operator Terminal Screen (POS); 3. Receipt Printer; 4. Barcode Reader; 5. Barrier Controller; 6. Alarm; 7. Document Viewer; 8. WIM Indicator.

b-Lane Equipment: 9. OHLS; 10. Fare Display; 11. Incident Capture Outdoor Camera; 12. Barrier Gate; 13. Traffic Light; 14. AVCC; 15. Vehicle Entry Loop; 16. Vehicle Exit Loop; 17. Outdoor ANPR Camera; 18. Smart Card or RFID Reader; 19. WIM Platform.

### *6.2.2 Plaza equipments*

The toll Plaza equipment consists of:

1. Database Server; 2. CCTV Display Screen; 3. POS Workstation; 4. CCTV (extra security) Cameras; 5. Report Printer; 6. IP Phone; 7. Network Switches; 8. IP phone Master (PABX) Unit; 9. UPS Power Supply.

## 6.3 Plaza network installation procedure

The first step, before the installation of any security system, is to ensure the presence of detailed drawings and documentation, a bill of materials with their specification. The installation of the ETC system uses many devices like vehicle-mounted electronic tags, toll point of sale(POS), RFID readers, and switches.

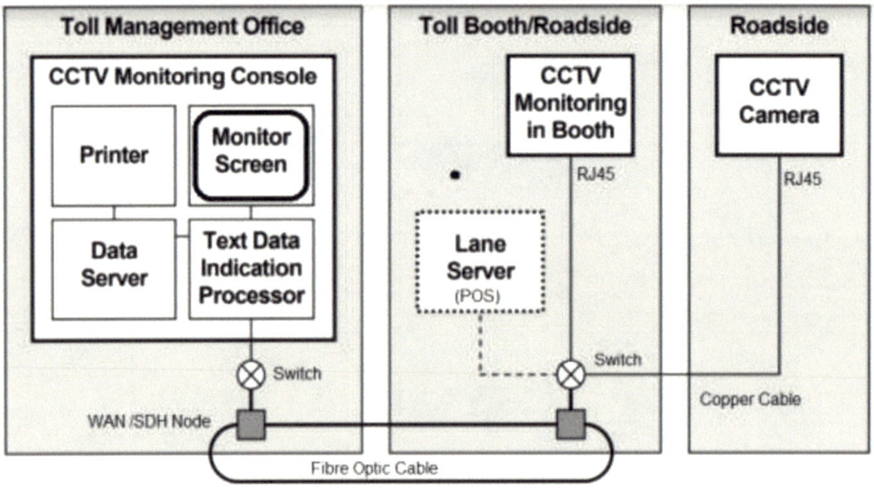


**Figure 10.**
*Connection scenario #1 of a toll plaza using a ring fiber between booths and plaza control room.*

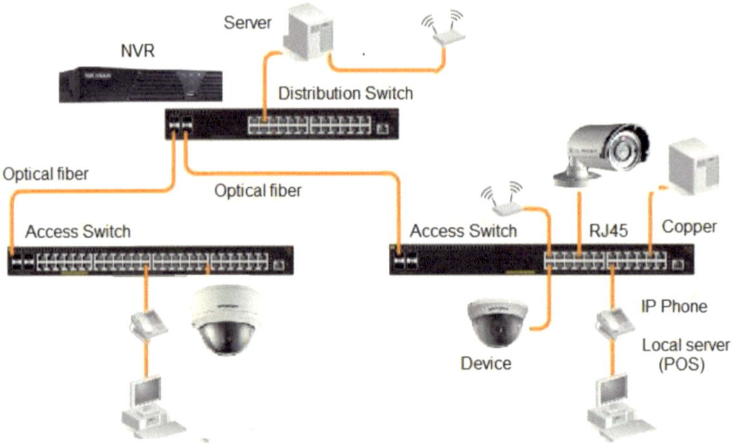


**Figure 11.**
*Illustrating example showing the internet.*

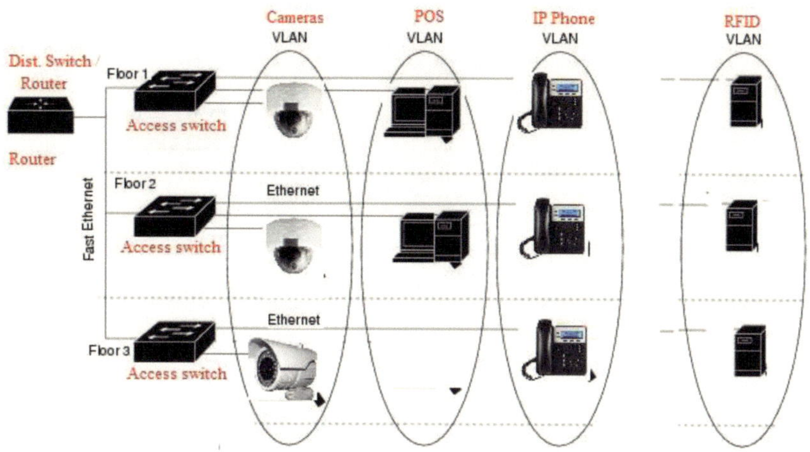


**Figure 12.**
*Example of a VLAN configuration of a toll plaza network.*

Actually, there exist several scenarios to connect the plaza network using Ethernet copper cables and/or optical fibers. The network topology may have several choices. For instance, you can use a bus topology, rings, μ-rings, or a hybrid bus/ring topology. One of these scenarios is depicted below in **Figure 10**.

If we have a sufficiently long multi-core optical fiber cable, you can also install a large ring between all switches in the main booth and the control room.

Note that the design of the toll plaza data network should be secured and isolated from direct exposure to other internet users. In particular, the sensitive video signal (from IP cameras) should be routed indirectly to the internet, via NVR, followed by a separate NIC card to the plaza server. The plaza server can be then connected to the internet by another NIC card. This is illustrated in the following **Figure 11**. In all cases, a virtual private network (VPN) should be installed before routing the plaza signals, to any external WAN, such as the internet.

In all cases, the data switches are dividing the core network into small subnets; called VLANs. For instance, by sorting node devices by functions or positions (camera, POS, etc., VLANs are often associated with IP subnets. Hence, networks with different VLANs will not be visible to each other. On the physical layer, the network remains the same as shown in **Figure 12**.

### 6.4 IP planning

The only way for someone to access the CCTV system is to know the IP address, username, and password (**Table 1**).

### 6.5 Port forwarding and accessing the internet

To remotely view a security CCTV system, you had to allow it to communicate to the internet. To allow access to your system from the internet, you have to configure the firewall inside the router to flow through the NVR. This process is called port forwarding and requires some advanced knowledge. There exist many guides on port forwarding if you feel confident configuring remote viewing. Each router will have its own method for port forwarding and I recommend checking PortForward.com for your specific model.

| Camera No | Camera name | IP address | Location | Username | Password |
|---|---|---|---|---|---|
| 1 | Front PTZ | 192.168.1.81 | Front | assigned by NVR | assigned by NVR |
| 2 | Back PTZ | 192.168.1.82 | Back | assigned by NVR | assigned by NVR |
| 3 | Booth Camera | 192.168.1.83 | Booth | assigned by NVR | assigned by NVR |
| : | : | : | : | : | : |

**Table 1.**
*Example of IP planning of the overall data network.*

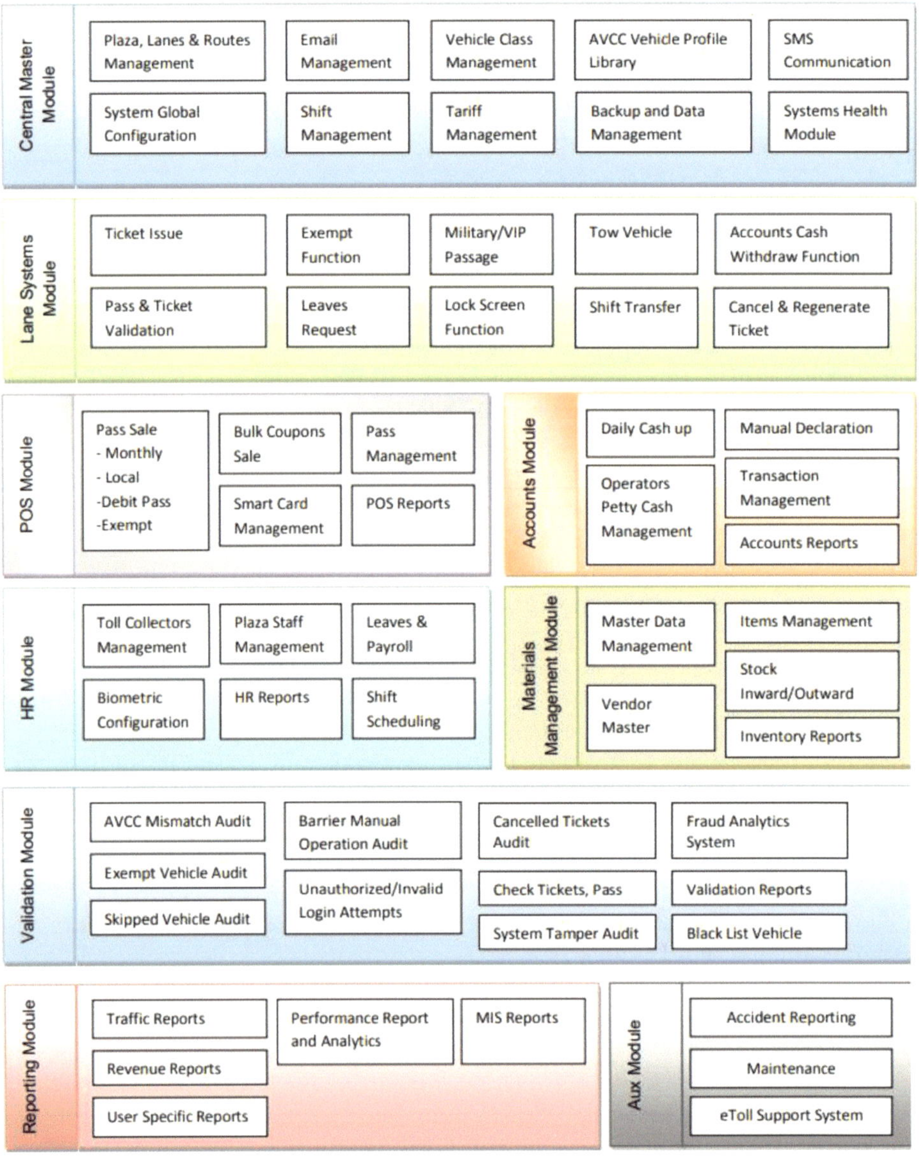


**Figure 13.**
*Modular architecture of Xanado VMS, with application to traffic surveillance and traffic control at a toll plaza network.*

### 6.6 Remote viewing using a smartphone

There exist Android and iPhone network operating systems (NoS) Apps that work with CCTV systems, such as IPTecno Pegaso [12]. This can connect to the system and display live video feeds. It also allows for one-way and two-way audio interaction, PTZ camera control, and motorized camera control. To learn how to connect your system, please follow any guide on how to view security cameras from iPhone or Android, such as this: https://www.cctvcameraworld.com/how-to-view-security-cameras-from-phone/.

### 6.7 Plaza database server and toll management system

The Smart Toll Collection System can greatly expedite the time taken by each vehicle to pay the toll fees. There are existing solutions that are deployed and are practically well-suited. A distributed database architecture platform should be installed to enable operations of ticket issues and validation. In the control system, the failure of the LAN connectivity should not impact the lane operation.

### 6.8 ETC management system

The following **Figure 13** depicts the modular architecture of Xanado, with emphasis on traffic surveillance control at toll plazas. Such software provides comprehensive capabilities to manage toll collection operations (24/7). The Central Administration Module facilitates the entire operation of monitoring and collection across toll plazas as one centralized unit. The Plaza Module can configure, manage all plazas toll collection operations, report toll collection in an audited manner. The Lanes Module is running under the plaza module, lanes module facilitates accurate toll collection for the vehicle passing from the toll plaza.

## 7. Conclusions

Public concern over security in recent years has driven the demand for video surveillance. Security and video surveillance need continuous change with time, to cope with the new hardware capabilities of IP cameras and video storage equipment. In addition, the increasing need for video analytics is required in modern VMS software, to perform their job of monitoring activities and protecting humans and their properties. Governments and police departments worldwide are constantly looking for new CCTV surveillance features that will help prevent crime. The latest VMS software relies on CCTV surveillance systems that can monitor a larger number of cameras and sites more efficiently. Therefore, combining SCADA features with VMS is significant.

This chapter presents a method based on neural networks (NN) for monitoring and operating video surveillance systems (VMS), like those in traffic control networks of electronic plaza sites. The method suggests that the thresholds used for generating alarms can be adapted to each surveillance device (e.g., IP Camera).

The industry needs to do more research on hybrid systems that combine the best of SCADA and AI algorithms together with VMS software.

## Author details

Muhammad H. El-Saba
Professor at the Dept of Electronics and Communication Engineering, Ain-Shams University, Engineering College, Cairo, Egypt

*Address all correspondence to: mhs1308@gmail.com

## References

[1] M. H. El-Saba, Supervisory Control And Data Acquisition (SCADA). 2017. Measurement & Instrumentation Systems, Publisher: Hakim. Available from: https://www.researchgate.net/publication/353142664_Supervisory_Control_And_Data_Acquisition_SCADA

[2] Mariana Hentea. 2008. Improving Security for SCADA Control Systems. Available from: https://www.researchgate.net/publication/253290135

[3] Marugán AP, Márquez FG. SCADA and Artificial Neural Networks for Maintenance Management. In: International Conf. on Management Science and Engineering Management. Cham: Springer; 2017. pp. 912-919

[4] El-Saba MH. Xanado, Video Management System, to be published

[5] Whittaker D. Why AI CCTV is the Future of Security and Surveillance in Public Spaces. Security magazine, USA, securitymagazine.com, 2021

[6] Bharadwaj HS, Biswas S, Ramakrishnan KR. A large scale dataset for classification of vehicles in urban traffic scenes. In: Proceedings of the Tenth Indian Conference on Computer Vision, Graphics and Image (ICGIP). Published in Association for Computing Machinery (ACM), New York, NY, USA. 2016

[7] Mohana et al. Design and implementation of object detection, tracking, counting and classification algorithms using artificial intelligence for automated video surveillance applications. In: Advanced Computing and Communication Society (ACCS), 24th annual International Conference on Advanced Computing and Communications (ADCOM-2018). Bangalore: IIITB; 2018

[8] Kain Z et al. Detecting abnormal events in university areas. In: 2018 International conference on Computer and Applications Beirut, Lebanon, IEEE. 2018

[9] Mohana, Aradhy HVR. Design object detection and tracking using deep learning and artificial intelligence for video surveillance applications. International Journal of Advanced Computer Science and Applications. 2019;**10**(12)

[10] Turaga SC, Murray JF, Jain V, Roth F, Helmstaedter M, Briggman K, et al. Convolutional networks can learn to generate affinity graphs for image segmentation. Neural Computation. 2010;**22**(2):511-538

[11] Balamurugan Rajagopal BGSA, Parasuram K. A robust framework to detect moving vehicles in different road conditions in India. Journal of Theoretical and Applied Information Technology. 2019;**96**(1):1-14

[12] Wang P, Li L, Jin Y, Wang G. Detection of unwanted traffic congestion based on existing surveillance system using in freeway via a CNN-architecture trafficnet, 13th IEEE Conference on Industrial Electronics and Applications (ICIEA), Wuhan IEEE. 2018. pp. 1134-1139